Engineering Management: An Irreverent Primer

Jack L. Wells

ISBN 145630075X

EAN 13: 9781456300753

Published December 2010

Sextant Publishing
a division of Wells Inspections, LLC

About the Author

Jack Wells enlisted in the U.S. Navy in June, 1960 right out of high school. He attended Penn State University (BSEE, minor in Management) under a full Navy scholarship, and was commissioned Ensign, USN, in 1966. He served aboard destroyers, intelligence ships, and mine sweepers in Japan and the war zone during the Vietnam War. Released from active duty as a Lieutenant in December, 1970, he progressed to Commander, Surface Warfare, in the reserve component, serving as Commanding Officer for four different units and retired with 23 years total service.

His civilian career started in 1971 as a maintenance general foreman at a food processing plant. He served as plant engineer for two sites. Relocating to the corporate office he was a project engineer, senior project engineer and manager of engineering. In 1981 he became director of engineering for a small international processing firm with two US plants. He constructed a meat processing facility in Venezuela, expanded fruit processing in Mexico, and dry powder handling in Canada.

In 1983 he joined a large multinational consumer products company, where he was a site engineering director, then a multi-plant engineering director, a regional technical director and ultimately a top corporate technical executive with responsibilities involving twenty plants and multiple engineering managers. He managed many very large projects and engineering teams. He was instrumental in the development of new products, new equipment and the construction of factories in the USA, Canada, Mexico, Brazil, Argentina, and Indonesia. He consulted extensively to business operations in Europe, Asia, and Australia.

He is now retired and lives in Florida, operates a commercial and home inspection business and has written and published three novels: *Quicksilver: a Greyhound at Sea*, a Vietnam war story, *Paper Dragon, Wooden Ship*, another Vietnam War story; and, his latest novel, *Breath of the Choson Dragon*, a story about current affairs and a confrontation between the USA and North Korea. His books have received multiple awards.

He and his wife Terry travel when they can and are members of the Tampa Sailing Squadron. They spent seven months aboard their sailboat cruising the islands of the Bahamas.

Preface

This book was written for the new engineering manager and for those who aspire to becoming an engineering manager. The step from senior engineer to manager is a big one and the management of other professionals is not easy. Engineering management is both technical and business management. It is a learned skill and every engineer so inclined can learn it.

This is my attempt to get the things I have learned over multiple years in engineering management on paper. Since I had a varied career from front line, hands in the gear, technician through all the intermediate steps to top engineering management for multiple businesses and industries, there are a lot of lessons learned. I also had the opportunity to attend multiple in house and external training programs. Each had value although some more than others.

Over the years I became convinced that engineering management was either not very well taught or not taught at all at engineering schools. I guess new engineering managers were supposed to pick up the additional skills required through osmosis like I did. It was a painful process and took a long time. Consequently, I made a concerted effort to train my engineering directors and managers as best I could. Now that I am retired, I can pass my approach to engineering management on to others.

Most of my personal experience has been in the continuous process industries. Food, both human and animal, was the majority of it. I have been involved in product, process, and project engineering and the management of those functions in everything from fruit, vegetables, sea food and meat, coffee, spices, baked and grain based products to confectionary. I have also been involved in the electronics industry and the manufacture of equipment to dispense food products. I was in responsible charge of the design, construction and start-up of multiple factories large and small in multiple countries.

Since I was fortunate enough to work for a company that valued technical development, I have been involved in the conceptualization and detailed engineering of high speed packaging and continuous processing equipment and systems. Most of these were unique and were developed to give the company a manufacturing competitive advantage – pushing the envelope beyond what was available commercially.

Over the years I have had the opportunity to tour factories that made airplanes, steel, wire products, automobiles, cigarettes, injection molded parts and films, paper products, chemicals, appliances, tires, automobile parts, tractors, floor tile, wall board, ceramics, recreational boats and even coast guard and navy ships. I have been in foundries, waste treatment and recovery facilities, shipyards and even oil and gas fields. I took every opportunity presented to see how the other half lived.

One thing always surprised me: there were a lot of differences but there were also a whole lot of similarities. Equipment was different, some larger like in a continuous casting steel mill and some smaller like in an alternator assembly line. But it all was very similar in concept and control. Some required large workforces. Some were so highly automated that one had to look carefully to even see a human. But the concepts were identical.

I have also had the opportunity to be involved with or tour construction sites worldwide. This included everything from high rise hotels to single story million square foot fabrication plants. Construction is generally generic. The end use of the structure and therefore its visible finish may be different, but the concepts are similar regionally. Construction contractors know how to build and don't care what the end use of the structure is.

So this book is not industry specific. Many examples provided are from the continuous process industry since that's my experience base. But the concepts are applicable to discrete fabrication also. It is a compilation of my years of experience in engineering management and my knowledge of what worked and what didn't. It is a "how to" book on engineering management from my perspective.

The first section is general management – and presented as a list of principles that I followed. Section II sets up some business measurements and project justification techniques. Section III gets into the nitty gritty. Section IV discusses some special situations. All are based on my experiences.

Hopefully this book will help the new engineering manager do a better job and not make the same mistakes I did. Make all new ones!

Jack Wells.

Engineering Management, an Irreverent Primer

Acknowledgements

Every successful person can, if honest with themselves, trace their success back to helpful individuals who contributed to that success. For me I guess it started with a demanding father who was also a carpenter. He taught me the value of education (he only went as far as 9th grade) and hard work. He also taught me how to work with my hands which, over the years, has been very valuable.

The US Navy was another stern taskmaster and experiencing the position of boot enlisted man convinced me that I did not want to spend long in such a situation. Then I became an officer and learned that the bottom of that ladder was not a nice place either. As I was told, there are Ensigns and whale excrement. Whale excrement is senior.

Once working as an engineer, I was lucky enough to have a mentor: the then manager of engineering. There is no substitute for a well connected mentor. There was a VP of Operations, who took a risk on a new engineering manager and taught me the business side. There was an operations director, who helped me integrate into a new environment. Later there were others, including a business owner, who really expanded my perspective with challenging assignments. I wish to thank these outstanding people for helping me grow. I don't want to name names, but if they read this they will know who they are.

Finally I want to thank my wife, Terry, who has suffered through the editing process both line by line and chapter by chapter for this and my previous books. There is nothing like having an experienced educator with a fine sense of spelling, grammar, and structure available to help an author turn rough drafts into useable prose.

Engineering Management, an Irreverent Primer

Table of Contents

Managing a technical function is not unique.

Technical competence is essential.

However, management itself is not technical.

Management is the application and coordination of resources to make things happen.

It is leadership.

And leadership is a learned skill.

-Jack Wells

Section I:
Basic Management Skills

Engineering Management, an Irreverent Primer

Section I, Chapter I

MANAGEMENT PRINCIPLES

These are my principles. They are a combination of my formal training and education subsequently developed over thirty years as I matriculated at the school of hard knocks. Some may be recognizable from other management texts although I sure can not remember where they came from. Some are totally internally generated. They are presented in what many would call an unorthodox, oversimplified, irreverent and directive style. Sorry. It is the concepts that count – not the delivery.

Effective management is the application and coordination of resources (time, money, and people) to accomplish organizational goals. It is a set of concepts applied consistently and repetitively over time.

Subordinates all learn at different rates and really want their bosses to be consistent, open to suggestions, respectful and fair. Managers who want their subordinates to do things one way today and another way tomorrow, whose moods change with the phase of the moon, drive subordinates crazy. Remember, consistency. Your subordinates want from you what you want from your own boss.

Every manager has a management "style". And you will develop your own. A "style" relates to how you plan, organize, follow-up, communicate and deal with your subordinates and others in your organization from top to bottom. My style would not work for you and yours would not work for me. Because it is all tied up in how you relate to your subordinates and others and how you have, over time, developed their respect for you. It involves your fairness and work ethic.

As a part of this, every manager develops a management "tool kit". Again, the "tools" are ways of dealing with situations that you acquire over time. You may have met managers whose only tool is a hammer. If you see all problems as a nail I guess that is all anyone needs. If it doesn't work, get a bigger hammer. Hammers leave a lot of marks on the surrounding woodwork. There are far more opportunities to use a jeweler's screw driver or a small paint brush to make fine corrections than there are to use a hammer.

Not that you do not need a hammer. Just use it seldom. Sometimes by letting your subordinates know that you have one and will use it if it becomes necessary is enough. And you need more than one hammer.

When I talk of a hammer, the ultimate hammer is firing: a good rap on the noggin with the big sledge hammer. Then there are a few taps with a tack hammer – like writing someone up, or sending them off to "discuss" their behavior with a human resource specialist. But in 90% of the cases you can just fine tune a subordinate's performance with precision small tools. The rule of thumb is to use the least invasive and least drastic tool to fine tune behavior early. Do not procrastinate. Un-dealt with behavior problems only get worse – never better - on their own.

Do not believe people who say a person is a born leader. There are some personal traits that may help, but good effective supervision is a skill learned through experience. Probably the single most important trait a person should have is a genuine respect for people. Given that, and maybe honesty and courage – leadership can be taught.

Throughout both my careers – military and civilian – I have come in contact with highly effective managers and highly ineffective managers. In most cases the difference was in how they dealt with people.

The good ones were perceptive, fair, open and consistent. A subordinate always knew where he or she stood with them. They rewarded exceptional results, accepted average performance, and dealt with nonperformance predictably. They were not screamers, volatile, or moody. They stood up for their people and developed their potential. They cared about the advancement of the organization and its success. They worked well on teams and communicated with superiors, peers and subordinates openly and truthfully. They understood that the success of the organization was because of the attitude of the people that staffed it.

They seemed to work under a set of their own guiding principles, and they shared these principles with their subordinates. These are mine:

Principle I: Managers do four primary things:

- Create and communicate visions (what does success look like, how do we set objectives and intermediate goals and plan to reach them)
- Establish priorities (what is the most important thing to do first ... second ...?)

- Allocate resources (time, money, people)
- Develop people (grow capabilities and prepare those with the aptitude for greater responsibilities.)

Only four, but good managers do them very well. They do some other things but these are the key ones. Let's expand on these concepts:

I (A): Create and share a vision of success with people. A shared vision guides and motivates far better than orders and directives.

A manager is the keeper of the vision for their part of the organization. This means that you get to interpret how your group fits into helping the organization achieve its goals. Get your subordinates to participate in this with you and get them to help you develop plans and goals for your group. They will feel more empowered, and they will help you make better plans.

Decide what success looks like and where you want to be at some point in the future. Plan long term, set goals. Act short term towards the goals.

Plans are not static. They need to be re-evaluated often and tweaked to meet the continuously changing environment. Plan, organize to achieve the plan, follow-up, evaluate, and re-plan. It is a never ending spiral and the more loops you make the better you get at it.

Some managers get their subordinates together for a planning session and start with a list of "We believes ..." Examples: "We believe that applied technology is a way to provide the business with a major manufacturing competitive advantage." "We believe that an internal strength of our group is our combined technical knowledge and our confidence in our ability to apply technology in new and unique ways." Then the group would get specific.

"We believe that we can develop a packaging machine that can fill bottles at 3000 per minute." "We believe that we can design a factory that has a net zero carbon footprint." "We believe that we can reduce product assembly time from a day to an hour." The above are all examples of developing an internal challenge and a vision. Then fill in with what intermediate steps are required and establish the "by when and with what resources." Like President Kennedy, build a vision and set challenging goals.

When President Kennedy said we will send a man to the moon and bring him back safely by the end of this decade, the technology to do so didn't exist. But he had faith in the fact that it could be developed.

As a side note, his "we will" speech was great for another reason: it was based on a vision and set clear goals. We will do this, by this date in the future and we will expend the resources to make it happen. It was a very clear, even if challenging, goal.

I (B): Establish Meaningful Priorities. The current business environment is fast paced. Upper management wants things now – not next week. And they all want it cheaper, faster, and perfect. That can't happen at the same time. You have to educate them.

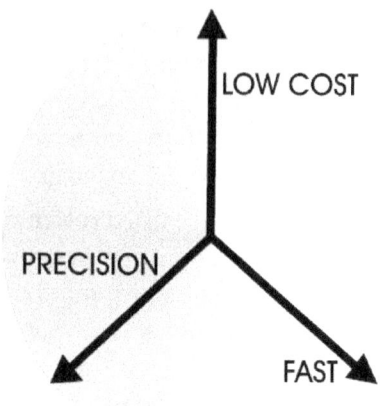

Almost anything a business does have tradeoffs. As the drawing shows, if you want it fast and precise, it isn't cheap. If you want it cheap and fast, you must give up precision. If you want it cheap and precise it is not fast. Most things a business does require knowledge of these tradeoffs and the selection of just what blend of fast, cheap and precision you can live with. No one gets all three.

Although some may not be able to articulate this, your people at some level understand it. They look to their boss to help them understand what the business needs and to select the right blend of speed, precision and cost that is obtainable.

Subordinates need to understand "why" something must be done, not just "what". This is setting priorities and helping them understand why you have set the priorities in the way you have.

This doesn't mean that you do not challenge them to do better. Set the goals high. Require your subordinates to stretch their abilities. Almost everyone can do more with less, do it faster, and be more precise than they think they can.

But if you expect your subordinates to do everything with nothing yesterday you are going to be disappointed.

I (C). Allocate resources appropriate to the vision and priorities you have set. Sounds simple? It isn't. You always have resource shortages. There is never as much time, money or human resources for everything you need or want to do to go around. Everyone in an organization is competing for the scarce resources. Budgets get cut and time scales get changed by internal or external influences. People get promoted, leave or are reassigned, leaving holes where they were before.

As a manager, your job is to do your best to insure that you allocate the scarce resources at your disposal to meet the priorities you have set. It is also your job to lobby for more resources and to convince upper management that if you get them, you can deliver and deliver more, quicker, and with more precision than your peers. Then go do it.

More resources sometimes require hiring new employees. Getting good ones is difficult. You will be selecting based on incomplete knowledge. What you see in a resume is what they want you to see. Resumes are notorious for embellishment and over statements if not down right lies. If you catch a lie don't hire the liar. Lying is a chronic affliction.

And watch out for titles. An Engineering Manager's job in a four person company is not the same as an Engineering Manager's job in a large multinational conglomerate. A flat organization may have titles that don't correlate to an organization with four levels of vice presidents and six levels of directors. Find out how many levels separated the senior business unit technical manager from the candidate's previous position.

A candidate's previous salary is a fairly good indication of level of responsibility. Their title may have been Engineering Manager but if their salary was $50K that is a whole lot different than an Engineering Manager who was getting paid $175K.

What you see in an interview is as good as it gets. Make sure your interviewing process includes some of your trusted subordinates and some human resource people. Go after the details. Not just what they did in a previous position, but with what resources, how long did it take and what were the technical and managerial challenges.

You should meet with the interviewers prior to an interview. Assign key elements for each to probe – technical knowledge, recent experience, breadth and scope of responsibilities, and human interaction. After the interviews, meet again to compare results.

Careful with the canned interview questions: "Where do you want to be in ten years? If you had a trait you would want to change, what would it be? What would you say was your greatest success – greatest failure/mistake? How do you feel about business travel, working long hours, working overseas?" Most candidates have already formulated answers to these that they think you want to hear.

In this day and age do not look down on candidates who are unemployed. But find out why. We all know that if a manager gets to select who he or she must lay off in a down turn that it is not his or her best people who get the pink slip.

However, mergers, bankruptcies, and factory closings do not discriminate well and the human resource attorneys will insist that layoffs go by seniority or some other system totally unrelated to performance to preclude law suits. That said, employed candidates looking for a step up say three to five years or less from their last promotion, are a good source of new hires.

Why three to five years? Well it takes most people three years to really get experienced in a job and to suffer the consequences of their own ideas. That is, to have the total experience of proposing an idea, selling it, executing it, and observing the final results. Less than three years and they are cleaning up after their predecessor and proposing their own ideas – but haven't had time to see the results yet. Much over five years and they could be stagnating and the good ones will be getting the "seven year itch" early and be looking for another challenge.

As a hiring manager I always took the candidate for a factory or engineering office tour. I insured that the tours included locations where something was happening. Naturally it was never in an area where a confidential and proprietary development was underway.

And I watched what they looked at and listened to what questions they asked. Technical people are, by their nature, curious. I have never met a good technical person who didn't enjoy a tour of a factory or technical office. It is an opportunity to learn. A candidate who shows little interest in what your business is doing should be sent a nice rejection letter.

Prior to making an offer to a candidate, contact all his or her references and his or her last employer. Do this every time and *without fail*. Attorneys have drafted rules for response to such a contact that may not give you a lot. Some companies will say that all they are authorized to do is acknowledge that the candidate worked there from date X to date Y. Try to probe deeper. Were they let go or did they leave on their own? Try to talk to their previous manager, not the human resource department. You will get a lot more insight.

If the candidate is currently employed, you can't call their current employer. Call the one previous to that. Right out of school? Call professors. Ask for a transcript.

If they state that they have this or that degree, have human resources call the educational institution and confirm. You would be surprised how many candidates claim degrees they just don't have or claim degrees from a diploma mill that advertizes in the back of airline magazines. Check a candidate's Facebook page or look them up on Twitter. You will gather some insight.

Hiring is easy. Firing a candidate who lied on his or her resume or in the interview is more difficult. And you look silly if you didn't follow up where possible.

Even with all the care noted above, remember the hiring rule: Even with years of experience in hiring, 25% of those you hire will be excellent, 50% average, and 25% you will have to, or at least should, let go within the first three years. Sorry. It doesn't get any better than that.

I (D). Develop people. If you make people feel respected and cared about, if you listen to people, they will deliver far more than their position would anticipate. In fact most, yourself included, can deliver far more than they or you might think. Challenge them and yourself to do more, faster, and with fewer resources. Genuinely respect people. Look out for their well-being.

It is amazing that some bosses feel that they must tell their subordinates everything to do and when and how to do it like they were robots. This makes your people feel small, resentful, and ultimately uncaring. But the single most important thing you want to do is to make them care about the organization and try to do whatever they can to make the organization achieve its goals.

Your management job is to make the people assigned to you grow each day in their job. Give them all the responsibility and authority you feel comfortable with. And the better the subordinate the more comfortable you will feel. Sometimes baby steps – later giant steps – but always make them feel that you respect their ability and give them challenges to expand their knowledge and skills.

Remember, the more you can delegate to your people, the more time you have to plan longer term – something essential for your organization and your success. It is called empowerment and it works!

You may run into the "that's not my job" people. I have little time for them but sometimes you have to deal with that type. The trick here is to tell them that no job in the business is static – it can not be for the business to grow. ALL jobs, your own included, expand over time and have to expand if the business is to grow and be successful.

So an item that may not have been "in your job description" yesterday may well be tomorrow. And written job descriptions capture a point in history – usually quite a while ago. Human resource groups use job descriptions to attempt to relate different jobs and set salary ranges. It is an inexact science. Job descriptions have little relation to a job today or tomorrow – like yesterday's news.

You need to periodically get out of your office and wander around, talking to your people, peers, and others and see what is going on. The view from your office is the view they want you to see. Take a walk now and then – unannounced.

Read the book the "One Minute Manager". It is old, very short, but very insightful. Originally published in 1982 and recently updated. There is an entire series of "One Minute ..." books but the original is the best from my perspective. Authors: Ken Blanchard and Spencer Johnson and available through major booksellers and online stores.

Employees want to feel that their boss is genuinely concerned for their welfare. Bosses show this by helping them grow, going to bat for them with the organization, and being firm but fair.

And every so often you find a subordinate that has demonstrated the potential for significantly more responsibility. Great; develop him or her and be their press agent. Sing their praises to your peers and superiors.

Employees who think they have the potential for promotion, and are correct in that assumption, "boss watch". That is, they watch their boss and the way he or she handles situations and makes decisions. And they form their own opinions as to how they would handle a situation if they were the boss. You can use this. Ask them how they would handle a situation. See how they are progressing. Maybe even accept that their solution may be better than your own.

Have you ever heard of a situation where a manager can't get promoted because he or she is too valuable where he or she is, that there is no one to take their place? Self inflicted wound. If that manager had selected and developed a subordinate to the point where they could take over, they wouldn't be in that fix.

Principle II: People need to be ready, willing, and able to perform. You can teach them to be ready. You can motivate them to be willing. If they are not able, find ones that are.

It is your responsibility to insure that your staff is capable of meeting the organization's needs. So if you have an employee that is not "ready" to perform because they lack some skills – you must teach them. Or at least arrange for them to get the training they need. This can happen by assigning them a mentor, sending them off to a training course, arranging for a six month cross assignment in another department, or encouraging attendance at a local university, etc.

You also must insure that their technical skills are up to date. Technology changes at a rapid rate. Keep your people up to date with both internal and external technical training. Otherwise you will get yesterday's solutions to tomorrow's problems. Not much of a competitive advantage for the business.

For newly minted managers who must supervise other technical professionals and interface with the business more, the American Management Association's four week (one week at a time spread over 12 to 18 months) Management Course is expensive but outstanding. It is very good at turning people into managers with a broader perspective; something essential for technical managers.

If you have an employee that is not motivated (willing) then you have to motivate them. Open your tool kit. Sit down with them and discuss motivation. Ask them why they appear unmotivated. Surprising? No, in many cases they will tell you flat out why. Or they will not and just tell you that they will try to do better. My experience says, "I'll try harder," usually doesn't work or at least work for very long.

If they really want to change, they will unless they have a personal problem. As noted later in this section, you don't do "personal problems". You get help from someone who is a professional in the field.

You should provide some time for a subordinate who needs help to obtain that help.

There are very few subordinates with motivational deficiencies that you can not improve. Sometimes just your attention works. Someone else cares. Subordinates want to be recognized for their contributions and to meet expectations.

Remember the very old Westinghouse experiment? Here the company had a group of employees in a room doing minor and repetitive assembly. Productivity was low. Management turned up the lighting. Productivity improved. They turned up the lighting again. Productivity improved again. Then they turned down the lighting one click. And did productivity drop? No, it improved again! Why? Because any change in lighting was seen by the employees as an indication that management cared about them and were interested in their productivity.

If you have an employee that is just not able to do his/her job – replace them immediately and get someone who is able. When I say not able I mean it's in their very make-up. They do not have the inquiring mind and technical aptitude to be an engineer. They can't interface with others without sparks. They can't manage money no matter how hard they try. An unable person will be an enormous drag on your entire group.

Sadly, there are people out there who have engineering degrees that should never be engineers. They got into it for the money. They were told that engineers make more. Smart enough to handle the academics; they ended up with a degree. But they can not apply the technical background they have to solving real world problems. These people need to be directed into careers where they will be successful.

Sometimes firing is also your job when it becomes apparent that a subordinate is not able to perform. This is the number one failing of most managers – they just do not want to fire an unable (or impossible to motivate) person.

But you must do it, even if in your organization you have to jump through a hundred hoops and walk barefoot over hot coals with performance reviews, write ups, etc. to finally make it happen. Sometimes in this process something will get through to the severely unmotivated person and a miracle will happen – they will see the light and start to perform. My experience says miracles happen very seldom.

Every organization has a list of "firing offenses". These range from stealing, making false reports, physical sexual harassment, fighting, and others. If one of your people commits such an offense you must execute the required consequence. Involve your human resources organization and do it as soon as you determine their guilt. Do it without fanfare or public humiliation. Attend them as they clean out their desk, collect their company ID card and credit cards, and walk them to the door. Insure security is informed that they are no longer employed at your facility.

There is a rule of thumb that says that ten percent of your people will require ninety percent of your time. This means that ten percent get all the attention. And ninety percent get short changed. Try to improve this. Your good people need encouragement. Don't spend all your time dealing with the non performers. Evaluate using the "ready willing and able" principle and take decisive action.

How about yourself? How do you stay ready, willing and able? I was always asked, "Should I get an MBA?" Answer: yes, if you want to climb the management ladder. An MBA will help you understand what makes business tick. I never got a formal MBA. But I had half of one when I graduated from five years equivalent in college with a minor in Management. I then educated myself extensively through self study and multiple off site courses. There are a few companies that encourage an MBA for an engineer to move up the management ladder and are willing to pay for it.

"Should I get an advanced degree in engineering?" Answer: maybe, especially if you work in government or research. All education has value, and not just formal education. To stay ready you need to keep your technical skills current enough to insure you understand developing trends and new material applications. To keep yourself willing, watch your attitude and remain positive. You are able or you wouldn't be reading this book.

"Should I become a Registered Professional Engineer?" Answer: yes. Do it as soon as you can after graduating from college and while you are working as an engineer prior to moving into engineering management. Keep it current and if you move from state to state get a reciprocal stamp. You may never have to stamp a drawing or set of specifications but having your PE registration gives you creditability, especially with government agencies and consulting engineering firms. Use your PE stamp sparingly. Never use it if it is not required.

Principle III: Treat your people fairly not equally. Equality is not fair. It rewards the non performers and short changes the performers.

Your responsibility as a manager is to fairly evaluate the capabilities and performance of your subordinates. You will be writing their performance reviews. And the review should fairly recognize the subordinate's contribution and personal growth throughout the reporting period – not just the last few weeks. But call a spade a spade. If you have an employee that is failing – do not wait until a performance review to spring your opinion on them. Do it as soon as you see the behavior that is not contributing.

If things do not improve you have a choice – you can gig a frog or you can kiss a frog. Sometimes, one you kiss turns into a prince or princess. Super when it happens. Usually all you get is a bad taste in your mouth.

Do not make a person your personal "project". But do try to find a spot where a non performer in one job may be a star in another. You may be surprised. Be objective – and fair, but not foolish.

Do not get tangled up with a subordinate's life. If their personal problems are adversely affecting their performance on the job, do not be afraid to discuss it with them. Then turn it over to the human resource professionals. You are not nor should you be an advisor on a subordinate's personal life.

Praise in public – chastise in private. You do not want to degrade a subordinate in front of a group of his/her peers. It creates embarrassment and deep resentment. If you need to chastise, take the offender aside, take a walk with them, or have a short discussion in private. Public praise for a job well done – a real job well done not some created praise opportunity – motivates all. Don't think your subordinates don't know when someone has done something exceedingly well. They watch their peers more than you do.

Be careful of making your performance discussions a ritual. If every time a one of your people is called to your office late Friday afternoon and you shut the door, he or she either gets fired or severely reprimanded, you can bet that the entire office knows this and watches the clock on Friday carefully. Someone called in then has to walk the "green mile" while their peers watch. Performance reviews do not need to be predictable.

Do not get talked into treating your people equally – and all human resources departments will try their best to get you to treat people equally because it makes their job easier and keeps the lawyers happy. But equal treatment is like communism. The real performers do not get recognized and the slackers get a free ride.

If one of your people comes up with a good idea do not take credit for it yourself – in fact praise him or her for the idea yourself and to your boss. Even better, praise him/her in front of your boss. You still get points for encouraging and developing your people to come up with good ideas and he/she gets an ego boost. Both help you and the organization advance.

Principle IV: You can not be a leader and still be close buddies with those you lead. Accept that command is a lonely place sometimes. You want to be respected, not necessarily liked.

Working closely with a subordinate is good – sometimes – but assigning them a responsibility and giving them the resources they need to fulfill that responsibility is more effective. You can not be their buddy. You may want them to like you – but not at the expense of thinking of you as a good friend. You can never be their good friend. You are their boss – inside the workplace or out side the work place. Have lunch with them now and then. Participate in a social engagement every so often – usually associated with work - but best friends: never.

This is the single largest mistake new supervisors make – they want to be buddies not bosses. You can not be someone's best friend and then write their performance review objectively.

Be friendly, open, share a laugh now and then, do a little socializing. But keep it to a minimum. Some employees think that if the boss wants to sit around and shoot the bull that they have to also. And they will, telling others behind the boss' back that they can not get their work done because they have to sit around and listen to their boss pontificate.

Minimize shooting the bull, and keep your people and yourself on task. Many good managers have established an informal time for socializing – like first thing in the morning, or over break, maybe right after lunch. People quickly learn the correct time for socializing by boss watching.

Do not be afraid of admitting your mistakes to your people and your boss. We all make mistakes. Learn from them. And making mistakes makes you human – something you need to be to your subordinates. Making a little mistake and getting laughed at is awful for some managers. But it works wonders for showing your people that you are human. Getting embarrassed, showing it, and laughing at yourself just makes you more respected.

Do not micromanage your people: it drives them nuts. There are lots of different ways to accomplish the same task. You can tell them what you want done, and why you want it done, but in most cases how is up to them unless the way is critical.

You can get a reputation as a perfectionist if you wish – but be objective – just how much perfection is required to give the same result? Remember, 100% is super, but life moves forward very well at the 95% level. And so will you. Your competition delivers at the 75% percent level. You then have a 20% advantage!

Do not talk down your bosses to your people. The "guess what the fat cats want us to do now" stuff shows you do not respect your bosses, and leaves the door open for your subordinates to disrespect you. You have to be able to tell your people what needs done and more importantly why. Worse case is to say, "I really do not understand why we are doing this yet, but someone thinks it is important and we will give it the professional try."

Principle V: Anger is never the right answer.

If you get angry, you loose control of yourself and consequently the situation. To remain objective you must control your emotions. Arguing is normally counter productive. Things get said that are not meant and feelings get hurt. Discuss openly and try to keep emotion out of it.

Sometimes you get trapped. Presenting a person with your opinion that their performance is not meeting expectations can trigger anger and even disbelief. This is especially true of an employee who has been around a long time and none of your predecessors has really broached the subject with the employee before. Have your facts at hand. Show them concrete examples. Don't use hearsay evidence. Even give them copies of the examples and agree to meet again in a few days to pick up the discussion.

Do not get into an arguing match. In most cases they will come back ready to talk turkey. Some will prepare a detailed defense. Evaluate the defense fairly. Your perception could be wrong. If your perception is based on what you have heard from others, not concrete examples, it very well could be wrong.

With the subordinate prepare a plan where they get to demonstrate, to your satisfaction, how they can meet your expectations.

Experienced managers have learned how to show displeasure without anger. Almost like the "I am disappointed in your behavior ..." thing parents do with children only more professional. Employees do not want disappointed bosses.

Yelling at people does little to change their behavior once they figure out that it is only noise. Like a threat never followed up on. You have to deal with them, not yell at them.

Never *ever* make a threat you are not willing to follow up on. If you tell someone that if they exhibit their uncooperative behavior again you will write them up – and they do it anyway then you *must* write them up. Threats that are not backed by actions are more than worthless – they actually reward the uncooperative behavior. The uncooperative subordinate has "gotten away" with something and in some groups that gives them status with their peers.

All that said, try not to use threats. They are negative. Establish clear guidelines. If someone violates one, they and all their peers should know for sure that there will be consequences and what those consequences are. Execute consequences with fair and solid judgment. But *always* execute them.

Principle VI: You have as much power as you are willing to exercise.

Most organizations suffer from a power vacuum as members try to keep from overstepping boundaries. Overstep. It is usually always better to beg for forgiveness than ask for permission.

The middle management supervisory staff is so afraid of taking on responsibilities and making decisions that the organization gets all tangled up with its own shoe strings and does not progress. Keep your boss informed, but do not assume you can't – assume you can, and then have a plan and then check with your boss. Most bosses are pleasantly surprised when a subordinate comes up with a plan of action to resolve a situation on their own.

Learn how to stage your ideas if you feel they will run into resistance if just presented to a cold audience. Like writing a mystery: you don't put the final scene in the first chapter. Sometimes you need to slowly build support for your ideas. You need to present your ideas in a logical stream of smaller but included ideas. Then, when you present a conclusion and the big idea, it seems like a logical progression of thought.

Principle VII: If it is not specifically precluded, it is allowed. If it is specifically precluded, it is negotiable.

Kind of a corollary to the overstep item above. Unless legally mandated, most managers have a lot more leeway than they know. Here again, human resource departments want everything structured with rule books (equal = not fair) and are masters of the "we do not do that here" line. If the rules are stopping you from achieving your group's and organization's goals – question the rules! That said never ever do anything illegal or unethical. It is your reputation at stake.

Rules in any organization are guidelines. If they are preventing the organization from making progress, they need to be questioned. Maybe they were created to prevent some behavior that caused grief in the far distant past. But are they still valid today? Has the world and the organization evolved sufficiently that the rule is overly restrictive now? Does it need to be re-evaluated?

In many organizations you can negotiate a dispensation from a rule if you can show how that dispensation will move the organization forward. Dispensation may only be a one time event since few senior managers will just kill a rule and may not even have authority to kill one. But many do have authority to bend or even break one if it is in the best interest of the organization.

Principle VIII: Being a good administrator lets you stay where you are. It does not get you promoted.

Lots of people want to do things right. A good manager and leader does the right things, and does them right the first time.

A managers' job is to do the right things – and sometimes to determine what the right things are. Yes, everyone has to fill out the government forms properly or to keep records filed for the period that the law demands. Everyone has to account for the resources allocated to them. But being a good clerk is not moving the organization or you forward.

Every organization has people that do the clerk function. It is generally a thankless and repetitive job. In some cases these people spend most of their day just doing what they have been taught to do. Maybe from the person they inherited their job from 15 years ago!

So they spend their day: Collect data from A and type up a report, copy and send the copies to B, file originals. Question? Are these reports still needed?

Do they serve any purpose at all any more? What does B do with them? Does anybody else ever read them or even know they are there?

You may be surprised. The clerk in department B may duplicate and forward to C, retaining a file copy. And C just files them in the circular file since no one ever told department C's clerk what to do with them. It's a whole lot of meaningless busywork. All these clerks would feel better about their job and make a greater contribution to the company if they had meaningful work to do.

Think of clerks as administrative assistants and use them accordingly. How much of the mundane day to day stuff you do could be done by your clerk with a little guidance from you? Wouldn't that free up your time for more important things?

You see that large wall full of file cabinets? What's in there? Does it need to be in there? Engineering should maintain a record of every project completed. You will need the information if there is a major disaster and you need to recover. Tax law requires records of assets that are being depreciated. But there are limits on how long other information needs to be retained.

How about that long row of book shelves stuffed full? If it has catalogs, just how up to date are they? Your engineers all use online catalogs now anyway, don't they?

Just how many years worth of Engineering News Record or Plant Engineering and Maintenance periodicals are enough? Could anyone even find anything they needed in them if they started looking?

Principle IX: Success is one part talent and nine parts persistence. There is nothing quite as common as unrealized potential.

Talent is wonderful. But by itself it is worthless. It takes practice and persistence. A persistent person will always win over the talented one over the long term. Said another way: recognize talent as a seed that if planted and watered, fertilized, and given the right amount of sunlight, may grow into a flower: talent needs nurturing. Sometimes talent needs someone saying "yes, you can, yes, you can" like a good sports coach.

Effective development skunk works are populated with talented people overseen by a good coach who understands that the development of new technology is an iterative process with many more failures than successes. Development is sometimes referred to as the bleeding edge of technology as compared to the leading edge. As discussed in later chapters of this book, the management of technology development efforts is not for the faint of heart.

Dogged persistence is the way to success. It took Thomas Edison 1000 tries to build a working light bulb. What if he had given up after his 990th failure?

Principle X: There are 24 hours in every day and 7 days in every week. Do not be afraid to use the time – but use it efficiently and effectively.

Even if your subordinate is paid by the "hour", in reality they, like you, are paid for what they accomplish – not for how long it takes them to accomplish it. Hourly pay is just a way to establish a wage rate that can be compared with wages at other locations. Try to get your subordinate to think in terms of what they accomplish that moves the group towards its goals.

If your subordinate is paid hourly, and if you need them to work longer, then overtime is required. Use overtime very sparingly. It is expensive, and it motivates a person to work slower on regular time so that they can get overtime.

Do not become the "workaholic". You need time off to regroup. This makes you more effective at work. Take your vacations; try to limit your workday to 9 or 10 hours. There will be times where 16+ hours a day is required – but these should be special circumstances like starting-up a new production line or dealing with a major disaster.

Normal workdays are not 12 hours long. If you really think you are efficient yet still can not get your work done in 10 or so hours, question your efficiency. Take a hard look at how you use your time and how well you are delegating. Take a time management class. See if you can shed some responsibilities.

Remember, workaholics have messed up home lives. Their significant others are on their case for never being home. If children are involved, spouses are living a single parent existence which increases the chance they will have problems. Home is a place workers should go to have their batteries recharged, not discharged further. People with messed up home lives are preoccupied with the stress and strains of their personal lives while at work. They have difficulty focusing.

That said, every manager has to cope with the trash compactor of trying to do their job well and trying to maintain a balance with their personal lives. The trick is to take full advantage of slow times at work to maximize time at home. Then when work heats up you can sell more time away from home for awhile. But it is not easy to maintain balance in the face of the big squeeze.

And do not get caught in the "I intend to be there before my boss and leave after my boss" stuff. Most bosses will think that you are in over your head – not that you have such a wonderful work ethic. Remember, you are getting paid for what you accomplish – not for how long it takes you.

If there is a problem at work and you get called, and you can not solve it over the phone or have doubts, go in. Call your boss to inform him/her when you are there and have scoped out the situation. Your estimated time of repair should be sensible and conservative. Don't tell your boss it is an easy fix unless you are 100% certain that it is. Think of how happy your boss will be when you call them earlier than he/she expected with good news.

In an age of staff reductions and layoffs, remember that the recognized contributors seldom find themselves unemployed. The operative word is recognized. If senior management is normally in at 7AM, try being in the lobby at 7AM to say good morning *now and then*. Try sharing your morning coffee with your boss when he or she has hers. All you are trying to do is to insure that upper management knows your face. Sadly, when decisions are made as to whom to layoff, faceless names get first priority.

And remember, those who are known because they are the ones who can be counted on to step up and take on a problem, a special project, or difficult task and then execute it well, become well known by name and face.

Principle XI: Establish contingency plans. Things will go wrong. Anticipate it. Change is normal. Nothing is static. Accept, even innovate change.

Try to develop various scenarios in your mind. What happens if this happens? What are the odds? What do we do then? What can we do now to preclude a bad scenario from developing?

Then when Mr. Murphy strikes, you already have thought about it and can propose or take actions to minimize his effect.

Change frightens. Our human makeup causes us to tighten up. Our fight or flight instincts kick in. But change is inevitable. Change helped humans evolve. It there had not been change, dinosaurs would still be running the earth and we would be nothing more than lunch. And change helps organizations evolve. Accept change for what it really is – an opportunity to evolve yourself and your organization.

Remember the old adage: "doing things the same way you always did and expecting different results is the definition of insanity." To get different results you have to change something.

Principle XII: Everything is connected to everything else. The world is like a balloon. Poke it in on one side and it bulges on the other. Anticipate the bulge.

Nothing ever happens that does not have consequences somewhere else; usually unanticipated consequences. Try to anticipate them.

If I increase funding for development one, it means I have to decrease funding for development two. Who are the champions of development two? How will they react? What does this do to the morale of those working on development three?

In any business entity, reorganization and the reallocation of resources always pokes the balloon. So where will it bulge? Reorganizations not only affect the site where they occur, they disrupt internal communications networks to other sites like slashing through a spider's web. Expect it to require time for these webs to be re-spun.

Some organizations seem to reorganize often. It becomes a way to cull out the dead wood that is more palatable to the human resource attorneys. Reorganizing a poor performer out of a job is far less personal than dealing with their performance issues. Too much of this is disruptive as the organization focuses inward and everyone goes through the anxiety of wondering when, and if, they will get shuffled off somewhere. And everyone knows that reorganizations usually require everyone to do more with less and take on additional responsibilities. Energy expended coping with reorganization is energy not expended improving your business's competitive position. If you have a say, try to minimum reorganization disruptions. They are a major poke at the balloon.

One senior leader sent a message back to headquarters complaining that every time his organization had gelled and became effective and successful someone reorganized them. Then things fell apart for awhile and competitors took advantage of his weaknesses. He pleaded to stop the reorganizations. The leader was Julius Caesar, then Roman General of Gaul, and Rome instigated reorganizations of his Legions had doomed his first attempt to expand the empire across the English Channel.

If you poke the market place balloon with a new product or other significant competitive action, do not be surprised if your competitors poke it back somewhere else. In fact, expect it.

Principle XIII: Trust your intuition. If it feels bad, it may well be.

Intuition is experience talking to you; your experience. The more experience you have the better your intuition is.

Don't just tighten up. Run down your uneasy feeling. Why does the situation make me nervous? Is it my internal fear of change or is there really some hidden potential for a problem to develop? What can I do to counter potential problems?

Risk is inherent in business. The trick is to manage the risk. And to manage it you must first understand it. You can not avoid it. And if you take no risks you are never going to move the organization forward. The return on a zero risk investment is zero.

Risk adverse managers may be appropriate for a few positions – like a bomb disposal squad or an insurance underwriter. But in most organizations they are roadblocks to growth and success. And they usually do not get promoted. Managers who take but manage risks well do get promoted.

Principle XIV: Find the cause of problems not the symptoms.

Most disasters start with a string of problems that set the stage. Then one small triggering event happens and a disaster ensues. Find all the underlying causes.

An example: A steam boiler melts down at a large food processing plant halting production and putting five hundred employees on the street for three weeks. The boiler was not just having a bad day. Six months prior to the disaster the boiler was in top condition and passed its annual inspection with flying colors.

Then top management demanded a reduction in operating costs. Budgets were cut 10%. But when the maintenance budget was cut, maintenance management just passed the cut equally to every area. So four months prior to the disaster, preventative maintenance of the boiler feed water controls and feed water quality testing were reduced.

A steam coil in a corn syrup tank developed a leak. When the tank was warm enough, the steam control valve closed and the weight of the tank's contents forced some syrup into the condensate return line through a defective trap. And no one was checking boiler feed water for trace sugar anymore.

The secondary low feed water alarm system gummed up and started to intermittently ding its alarm bell. After repeated and un-responded to requests for maintenance help, one new boiler operator just stuck a rag under the ringer. Now the secondary alarm was inoperative.

Sticky sediment formed in the feed water level probe chamber and the float stuck. At normal steam flow rates the situation was static. But the stage was set.

As the factory started to ramp up a larger steam consuming process, the boiler responded by going to high fire rate. The float was stuck so there was no signal to the feed water supply valve to open more, so feed water supply couldn't meet demand. The boiler had a big fire and insufficient water. Tubes overheated and started to warp. It wasn't noticed until someone saw black smoke belching out of the stack that normally wisped almost clear smoke. The interior boiler tubes were burning. Lucky the boiler didn't explode.

The root cause was not the sticking feed water control. It was insufficient maintenance and irresponsible maintenance management.

These kinds of situations happen often. And a good manager will move to preclude them in advance. Here, something as simple as insisting on preventative maintenance to critical equipment would have precluded the problem. Yes, the budget was cut. But performing effective triage would have identified boiler feed water controls and tests as one of those things that could not be short changed.

Most disasters happen this way. You can go back and find a long chain of links that had to be in place before one small triggering event – the last link in the chain – caused the entire system to self destruct. If you want to preclude disasters you have to take action early.

Principle XV: Precluding problems is a manager's job. It is called forehandedness – the ability to anticipate. Eternal vigilance is the price of safety.

This principle is a corollary to the one above. The single most respected trait a manager can have is forehandedness. Their group seems to never have the "awh craps" that happen in other groups. They seem to have a sixth sense about averting problems and about how to move forward.

These managers maintain a healthy skepticism. They constantly evaluate based on measuring the potential for problems, the probability of their occurrence, and the magnitude of the disaster that might ensue. Potential multiplied times probability times magnitude gives you a number. Work on the higher numbers first.

It's like dealing with a fire. Fire needs three things: fuel, oxygen, and initiating heat. Take any one away and there is no fire.

It would do every engineering manager good to read the final reports for the 1986 Challenger shuttle explosion disaster, the 2003 Columbia re-entry burn up disaster, and the 2010 Deep Horizon oil well blowout disaster. In each case the controlling organization went from, "prove it is safe," to, "prove it is unsafe," due to political or economic pressures. This mind set allowed multiple rationalized mistakes to be made and a chain of individual, "we can squeak by with this," problems to form one link at a time. Proving that individual small deviations from standard are unsafe is very difficult. So the chain grew and the last, "squeak by," link triggered a cataclysmic disaster. Never, ever, relinquish the requirement to prove that the entire system is safe.

Principle XVI: Discrimination is not to be tolerated.

In this day and age discrimination in the workplace based on age, race, religion, gender, or sexual orientation, etc. is wrong and an opportunity for a lawsuit. You must make it apparent to all your subordinates that you will not tolerate it. Slight offenders will be counseled and severe offenders terminated.

You may know of an older man that gets away with flirting and teasing younger female employees in a way that no one else could. They think of him as a harmless teddy bear. And he fosters that image. It is a time bomb. All that is needed is for one female to take offense and you have a detonation. Counsel him.

If you have to do a difficult performance review or counseling with a subordinate of a different gender or ethnic group, take appropriate steps to do it sensibly. Have a human resources person, maybe of the same gender or ethnic group as your subordinate, in the room with you as an observer. Have them document what is said.

Reverse discrimination is discrimination. Just because someone in the past was discriminated against does not give them the right for preferential treatment now. People should get preferential treatment for excellent performance, not for what group they were born into or raised in. And their pay and opportunities for advancement should be no different than anyone else.

Affirmative action may be appropriate to bring fairness back into workplaces where previous discrimination was widespread and systemic. I have never seen this in an engineering organization. The work was highly technical and minorities were valued for their education and experience. There were never enough people to go around and everyone generally had to be treated fairly to get the work done.

Human resource departments who are being evaluated against quotas or who have been directed to increase the number of certain groups in positions of responsibility will push you to be preferential towards minorities. Preferential treatment will sour the morale of your workgroup just as fast as discrimination will sour morale.

You may find that an individual with a different background may need some special counseling or training. That's fine. You treat your people fairly as individuals and develop them as individuals. Your human resources department has resources to provide this counseling or training. This is not preferential treatment, it is just common sense. You are developing your workforce.

Your entire workforce knows what is fair and what is not. If you have been given a subordinate that is a member of a minority group, expect the same performance from them you expect from all and give them the same encouragement you give to all. They will appreciate it. And if a member of a minority group is not performing, deal with them the same way as you would any other subordinate: firmly and fairly.

Principle XVII: Do not try to "wing it" with superiors. If you do not know say so. You get one opportunity to establish creditability.

If you do not know an answer when questioned the old adage, "I do not know right now, but I will find out and get back to you", works well. Then follow-up right away. Anticipate questions. Always there will be, "Why not cheaper? "Why not faster?" Have an answer prepared. Discuss alternatives you considered and why you discarded them. This builds trust. And trust builds power.

Unsubstantiated facts are the kiss of death. Many managers, especially the more senior ones, can spot an unsubstantiated fact a mile away and will grill the presenter unmercifully. If the presenter gets flustered and responds with even more unsubstantiation it can turn into a blood bath. The presenter ends up looking like a fool, as does his boss for putting him up there on the firing line, and creditability is forever lost. The presenter's boss will vow to never let the presenter walk up to a podium or pick up a pointing device again.

Principle XVIII: It is always better to make a decision, even if it is wrong, than to allow fate to make the decision for you while you procrastinate.

Fate makes lousy decisions because fate knows less about the problem than you do. No one gets to make decisions based on perfect knowledge. There are always some unknowns. In most cases you have a 50/50 or better chance of being right. The more experience you have the higher your chance of success. But if you procrastinate ultimately fate will make the decision for you. And fate is wrong 80% of the time: A bad decision. Most managers will give their subordinates more hell for procrastinating than they will for a wrong decision.

Snap decisions that have not been well thought out in advance are also usually wrong. Only skilled gunslingers can shoot accurately from the hip. And if they miss just once, they die. Take as much time as you have to think it through. Better still think through upcoming decisions in advance. Then if you really have to shoot from the hip your aim is better.

A few years ago there was a management philosophy touted called "ready, fire, aim". Some managers took this literally. Not a good idea. All the ready, fire, aim philosophy was attempting to do was get management to move things forward more rapidly. To aim a little, take a shot, correct the aim based on the observed result, and take another shot. Walk the rounds into the target with small adjustments.

Explained this way ready, fire, aim is a good way to take out a target quickly. Watch for collateral damage from the first few shots fired. It should have been "ready, aim a bit, fire, correct aim, fire …"

Principle XIX: A person's authority may come from a superior, but power comes from a person's subordinates: The power to make things happen.

Authority without power is worthless. You need power to make things happen. If you don't have it, soon the authority gets taken away. Take care of your people and they will give you the power – by working for and with you to make things happen for the organization. And the more empowered you make your subordinates, the more power they give you.

And they give you time – the manager's own most important resource. The more you can delegate the more time you have to plan ahead.

Remember that responsibility is unique. You can delegate responsibility, but you still have as much as you started out with – like an amoeba cell splitting. You retain not only the original responsibility but also now you are responsible for what your subordinate does with his/her new responsibility. And with delegating responsibility you must also delegate the right amount of authority to go with it. Responsibility with out authority is a guarantee of failure. Ever tried pushing a rope?

Most management jobs are split up into 3 parts: job related, boss assigned, and self assigned. A good balance is about a third each. With a new job, until you get to know it, it will be job 70%, boss 30%, self 0%. But ultimately you want to get to the 1/3 each rule. You need your 1/3 to come up with and do the things that will move your group and the organization forward over the long term.

The higher up you are in management the further out you should be looking and planning. A rule of thumb for front line supervisors is to be thinking about what your group should be doing next month. And planning and organizing for it. Day to day your people should follow your previous plan and all you should have to do is follow up, fine tune, and deal with contingencies.

If you come into a new job that is not organized you may well have to get into the day to day a bit at first – but remember your goal. Get to thinking ahead, maybe a week, then two weeks, and finally the month. The further up the ladder you go the further out you have to think. Top management has to plan ahead years.

Principle XX: Make your people function as a team. Groups working together develop better ideas and answers and you do not know it all nor will you ever.

Communicate and collaborate. If you do not have multiple subordinates, then you and your subordinate are a team – a team with a leader – you! Unless you just did their job for years, you will never know more about the details of your peoples' jobs then they do. Never. And you do not want or need to know.

But this means that you have to ask them if you really need to know a detail.

Recently there has been a lot of management theories espoused in literature and even tried in real life concerning something called a self managed group. A leaderless self managed group. These groups can and do make progress if their scope is very limited: "All second shift operators in the number two process area will meet to discuss how to reduce scrap" would be an example. One or two meetings will be productive if the group is given clear concise goals and has a precise understanding of what they are to deliver. Like a brainstorming session.

There is also a system called area management groups. Here people from both the line and supporting staff departments meet to coordinate activities; again productive for coordinating.

However, most groups do not perform well long term without a leader. Some leaderless groups create a defacto leader even if they call that person a facilitator. Call it what you will, effective groups need someone to call meetings, set agendas in consultation with group members and trigger or cut off debate. They need someone to insist that commitments are met.

Most of the highly touted, leaderless, self managed groups in industry have, by necessity, reverted to traditional groups with an appointed leader. Anarchy doesn't work for long in a business environment.

Principle XXI: Be firm and consistent. You get exactly the behavior from others that you are willing to put up with.

Never abide self serving people who do things that make them look good at the expense of others or at the expense of achieving the organization's goals. You are responsible for your group's performance and are judged by how your group performs. Do not allow a surly, uncooperative, or lazy person to continue that behavior. A bad apple can spoil the whole barrel.

If you put up with uncooperative or insubordinate behavior you not only look weak but you encourage that behavior. Gross insubordination is mutiny. Your entire organization will come unglued and slip into chaos. Counsel the mutineers and if that doesn't work hang them from the starboard yardarm. Never, ever, abide insubordination.

And do not "transfer" a problem employee to another group. If you got him/her then you fix it if possible or fire them if not possible.

People do things for a reason. Find out what their "currency" is and use it. Every change in anyone's behavior is driven by motivation. And motivators are not just the "carrot and the stick". There should be rewards for exceptional behavior, positive consequences from good behavior and negative consequences from bad behavior. Do not just use rewards or positive or negative – use all three and make sure they understand all three consequences clearly in advance. In simple terms – do what you are supposed to do and break even. Do the exceptional and get rewarded. Do something bad and get punished.

All this sounds like a prescription for being labeled a hard nose or tyrant. Please don't read it that way. Firm, fair, consistent is what you should strive for.

Managers that are real tyrants find that their organization quickly learns to tell the manager only what they think he or she wants to hear. They start to do exactly what the manager orders *even if it doesn't make sense!* They start to follow the letter of your directives, not the spirit of them and a quick response to a situation from you gets turned into action without question: a prescription for disaster.

Example: I was a brand new Director of Engineering in a medium sized company. Reviewing drawings for a new piece of equipment with the design manager, I commented that the huge size E drawings he was using were almost impossible to handle or fold let alone ship overseas to our subsidiaries. Not remembering all the standard drawing size lettered designations, I just commented that we should limit our drawing size to something like two feet by three feet in the future. He took me absolutely literally and had special drawing paper made up exactly 24 inches by 36 inches instead of converting to a standard size D (22" x 34")! Naturally, the new drawing paper couldn't be folded to 8.5" x 11" and folded wouldn't fit in an envelope or file. It was a teachable moment. Once I became aware I fixed it, but I learned something.

You want your people to feel totally comfortable telling you the truth and questioning an order if it doesn't make sense to them. Sometimes you have to draw the timid ones out with questions like "Do you understand what we need to do? Why we need to do it? Do you have another idea? Anything I missed here?"

Principle XXII: Never ask your people to do something you would not do yourself. And sometimes you must do it yourself to prove it. Lead from the front, not the rear.

If you ask your employee to work late on a special project, then stay there with them and even run out for coffee for the both of you. If there is a real shitty job that you would not do – do not ask your subordinate to do it. Do it yourself or at least side by side with them.

I'm not saying you have to climb into the sewer with a sewer cleaner. But you could show up to supervise in old clothing and tend the rope at the manhole. Show the sewer cleaner that you value the work being performed.

Recently there have been some television productions where a Chief Executive Officer of a large company dresses down and gets hired as a new employee at a remote site at the bottom of his/her organization. Usually the CEO is shocked to see just how hard the people at the bottom of his organization work for little pay and over long and shifting hours. They are surprised at the dedication and company spirit demonstrated by some employees. And they return to the executive suite wiser and more appreciative of the immense effort required to keep their company profitable.

Every now and then try to put yourself, at least mentally, in the shoes of those at the bottom of your organization doing the grunt work essential to keep your organization functioning.

If there is a problem in the group you are responsible for, it is your problem. Accept responsibility and then deal with it. We all know that in organizations excrement flows down hill. If it is headed for your group step forward and try to stop it. Dodging it or worst case adding to it is cowardice. Your people will never follow you in quite the same way if you run under fire.

If a disaster occurs at a site you are responsible for you must take charge. Get the local engineers to do what ever short term mitigation they can and get on the next airplane. You can not make good decisions remotely and the locals may well be shell shocked.

Most businesses can not cope with a production facility being out of commission for long. Product deliveries slide and lost sales may never be recoverable. You must get a handle on the cost of lost sales and then you will have a handle on what you can spend in an emergency to recover. Then mobilize your resources and get cracking.

Remember, in any disaster perception becomes reality quickly. A less than instant response is immediately assumed to be an uncaring attitude. Remember hurricane Katrina.

Example: Once one of our large factories experienced what is called technically an "arcing fault burn down". Heavy summer evening thunderstorms injected a huge electrical surge into the multiple wire aerial supply lines from the main transformer bank to the service entrance. It was an older factory and there were problems with the design of the internal power distribution systems. Essentially there was no ground fault protection. Electrical currents surged through the system building voltages until massive arcs occurred.

Most of the older 1600 ampere distribution breakers never tripped since the line to line fault never reached the 9600 amperes required to make them activate. Every large buss duct failed and buss bars looked like they had smallpox. Ultimately the main transformer bank's primary fuses stopped the damage, but not until one of the transformers faulted internally. Luckily, there was no large fire and no employee injuries, and the smaller electrical insulation fires were quickly extinguished. Ten minutes after the surge started the factory was dark and the entire electrical distribution system was reduced to junk and puddles of melted copper.

But there were five million gallons of fruit juice in cold storage and refrigeration was out. If the juice temperature got above thirty two degrees the juice would start fermenting and be un-saleable. It was a certifiable disaster.

As a brand new engineering manager, I arrived early the next day. First order of business was to meet with local management, tour the site, and get some contractors in. We started purchasing 500 mcm copper cable reels and running temporary power while the utility replaced the burned out transformer. We worked ourselves and the electrician crews 24/7. We used rented generators until we got the transformers back on line. In three days we had the refrigeration system back in operation. In two weeks we had the factory back in production although we had bought all the 500 mcm cable available in a three state area.

The factory operated with the temporary heavy wires everywhere until we could design and install a new distribution system. Of course the new system used impedance grounding for coordinated ground fault protection and our two other older factories were retrofitted with the same over the next few years.

Total site cost $3.5MM for temporary and $5MM for a permanent fix. But we saved $10MM in raw material and $30MM in sales.

Principle XXIII: Never be a "we can't do it" person. Keep a can do attitude. "Yes we can – but it may be difficult". Anything can be accomplished with the right resources.

Negative managers do not get promoted. In fact, some do not even get invited to meetings anymore since their attitude is considered to be toxic. Then they do not know what is going on. And knowledge is power. A manager without power ceases to be effective and ultimately gets to "explore other opportunities".

A positive attitude permeates the organization in general and your group of subordinates in particular. All subordinates boss watch. And many emulate. If your attitude was good enough to get you promoted than it ought to help them get promoted.

Sometimes your work environment becomes toxic to itself. People working in locations that have a distinctive "bad economic environment" get bitter. They see their jobs slowly being relocated to areas of the country or world with better business environments. They feel un-appreciated and afraid. But business is not static.

If you work in the northeastern or north central USA where taxes, energy costs, labor costs and etc. are high, your firm will ultimately try to find a location where things are cheaper. To a certain extent this is also happening on the west coast, especially California. Whining about how your location is getting "screwed" by the big bosses or the business is a real good way to lose personally.

Be prepared to move where the work is. In the meantime, try to make your location more valuable to the company as a knowledge base, flexible small lot manufacturing facility, development center, and etc. Whining about production going elsewhere for economic reasons makes you a liability.

Face it, businesses are economic entities. They make economic decisions or they perish. There may be some corporate loyalty to an original location in the high rent district. But loyalty starts to evaporate if competitors can put an equivalent product on their customer's table cheaper than you can.

I used to say, when asked if we could do something really weird and expensive, that yes we could. We could build them a Taj Majal, complete with reflecting pool if that is what they really wanted. It would take awhile and cost a bundle, but we could do it. If what they wanted would require a lot of technology development, then I would tell them that and fairly explain the cost and timing. Almost always they accepted the answer and decided if they really wanted it bad enough to try to sell the cost and timing upstream. Sometimes they didn't. I never ever said, "That's impossible," since few things really are if you have the time, money and manpower to make the attempt.

In most businesses it is far easier and less risky to kill an idea than it is to nurture one. And a weird idea today may be a brilliant idea tomorrow. Give the weirdos a chance to develop a little. Evaluate carefully before you summarily execute an idea.

Principle XXIV: Meetings can be the greatest waste of time or the most effective way to make important decisions. But they must be managed meetings.

Rules for meetings:

1. Preset agenda, copy for all attendees – "if we do this we will be done". And do not let someone take over and bring other items to the agenda. Set up a separate meeting to deal with a new critical issue.

2. Everyone arrives on time. If not the door gets locked. Another method, late comers put a dollar into a can for every minute they are late. Cans are collected and the money used for a division party at year end. Harsh but effective. Just think of the time wasted while a group sits around and waits for a late attendee.

3. Everyone at the meeting representing another department arrives with the power and authority to commit that department to a course of action if necessary at the meeting – no "I need to check with my boss on that one", stuff. If they don't have the

power to make a commitment, then maybe their boss should have attended the meeting.

4. Do not allow subordinates in the same meeting with their boss to take the floor and lead discussions – subordinates tend to want to talk to impress their boss not to move the meeting forward. Subordinates can be in the second row and available for consultation – but they do not get a vote or get to pontificate.

5. Get every major attendees input. If they don't offer it, pull it out of them. If they weren't knowledgeable enough to have an opinion they shouldn't be there in the first place. Remember, "I don't really care one way or the other," is an opinion. Respect it.

6. Once a decision has been made, poll all the principles around the table, "Do you agree with this decision – or at least can you live with it and support it?" Someone who says yes can't very well say they do not support the decision afterwards.

7. Then ALWAYS make up an action plan. *Who will do what by when* in support of the decision? Every assignment must include a what, and most importantly, *by when* statement. A decision without an action plan is worthless. Write the action plan on the board then write it in the minutes.

8. Every meeting worth calling requires outline minutes typed and sent or emailed to all primary attendees.

9. Then set the time and date for the next follow up meeting and tell everyone who has things to get accomplished that you will ask them for an update

on their progress at the start of the next meeting. Then do it.

10. If the meeting starts to drag – take a 10 minute break. Same rules – late and the door is locked. Any meeting on one subject that lasts longer than two hours is doomed to failure. You have too many attendees, an unclear agenda, an incompetent facilitator, a poor understanding of the problem or decision you wanted to address, or a room full of professional meeting attendees. People just tune out. After capturing the progress you made, adjourn it until you can get your act together.

Principle XXV: Be careful of tradition.

Tradition usually means that is how we always did things. And, as we said before, doing things the same way you always did and expecting different results is insanity.

Most successful managers view tradition as a not so sacred cow to be slaughtered and made into meatloaf for the cafeteria. Tradition worked well for the Catholic Church: Gave them total control. The Dark Ages were created by insisting everyone follow strict tradition. They only lasted 1000 years. In business most traditions, if they are strictly part of the culture, are Ok. Like a Christmas turkey for each employee, a summer management/hourly softball game, etc.

Traditions do give an organization an identity. The military services know this and foster many traditions. They are connections with the past and help maintain the pride of a group. But if they hamper the organization's ability to be responsive and dynamic, they need to go away.

Principle XXVI: History teaches.

Never make the same mistake twice. Make all new ones. Do not let history control you. You can not change it. Learn from it. Then get on with the future. Living in the past is worthless.

Wringing your hands in anguish does little to solve problems. If something goes wrong, the best course is to learn from it. If necessary make the changes in procedures and policies which will keep you from getting an instant replay. But don't change things if the problem was caused by someone not following existing procedures and policies. Evaluate why and take appropriate action.

In some organizations every unhappy event generates another multipage instruction, rule or policy that gets promulgated throughout. It is like the person developing these is being judged on the net volume of paper used. As time goes by these instructions fill filing cabinets and never get read. If the crux of a new policy can't fit on one page, it is too complicated.

Once we were sent a set of corporate finance manuals when we were just four people operating out of a tiny office on a remote site trying to establish a little business and build a tiny plant in a developing country. We didn't even have a finance person. Stacked up these manuals were six feet high. They made nice stools and foot rests.

Watch out for people who live in the past and always have stories about how they were discriminated against, punished for the transgressions of others, punished far in excess of their own transgressions, unfairly overlooked for promotion, or generally trod upon by the organization. And every organization has these people. They have not learned from history and have not taken the actions required to insure that they don't repeat it.

When you get appointed as a new manager you will get to hear all about these. Employees feel that new managers are fresh meat to air their disappointments with. They will also want to re-present their past ideas that were not considered favorably by their previous manager. Expect it and pay attention. A previously rejected idea may now be a good one.

In some organizations there is a relatively set career path. Some would call it filling out their dance card. In other businesses, the progression path is not so clearly defined and the path changes. Yet the history dwellers either do not understand the path, or choose not to. A prime example is those who choose to ignore the fact that large multisite businesses value experience at other sites as a requirement for promotion to levels of management that must deal with multiple sites.

The history dwellers refuse to relocate, yet get bent when others who have get promoted.

Most people do not understand the promotion process. People get promoted because they have the experience, attitude, technical skills and attributes that will fit and complement the peers they will be working with in their new position. Having just the experience, attitude, and technical skills gets a person considered – but not necessarily promoted.

It is equal to standing on a subway platform waiting for a train. If you have worked your way to the front of the line where the car doors will be, and if when the train comes in the door in front of you opens, you get to take the step aboard (get promoted). If you are not at the front of the line, if the train hasn't arrived, or if the door in front of you doesn't open – you get to wait for the next train.

Witch hunts are one of the most destructive behaviors that an organization can get into. Anyone even remotely involved goes into "cover your butt" mode. Progress grinds to a stop. Cliques form for self preservation like elephants forming a circle to provide protection from marauding lions. Finger pointing and outright falsehoods make investigations difficult. In severe cases the organization learns little from the failure except how to dodge artfully when the excrement hits the air handler.

Some organizations will require that one or more perpetrators be identified and publicly punished – mostly without anything approaching a fair and impartial trial. Locating the perps and bringing them to justice is not the number one goal after a problem pops up. Investigating the situation and identifying the root cause is far more important.

In a political arena, like in government, a witch hunt is pursued to make the opposition look bad. Ever watch congressional committees try to pin the tail on the donkey after something goes wrong? Does the real donkey ever get the pin in the butt? Seldom. Usually a sacrificial lamb gets presented and then placed on the altar with much pomp and applause.

If you must give up a sacrificial lamb, make sure it is the real responsible person – not a token to the gods. People in most organizations know full well who screwed up and how badly. If someone else gets the ax it makes the entire organization uptight and risk adverse.

Principle XXVII: Understand and use your resources.

Money is a resource, so are your subordinates. And *time is a resource*. You can always get more money and maybe even more people. Time is fixed. Once spent you don't get it back. Spend time very wisely.

If you spend your day at work jaw boning with others, discussing sports, and generally being unproductive you are wasting a major resource and your bosses will see it and resent it. Spend your time at work – working.

The personal computer is the greatest productivity tool ever invented. Knowing how to harness its power to improve your productivity is essential. Using it to model reality allows you to test ideas and helps you make better decisions. Building models teaches you how things interrelate. But don't climb into the box on your desk and tune out the world like your latest super spreadsheet or data base creation was a video game. A computer at work is a tool, not an entertainment device. The same is true of your PDA or iPhone. As a manager your job is to manage people. You can't do that buried in an electronic device all the time.

Email is essential for communications across organizations. Hopefully your organization has rules to minimize internal spam and filters to keep external spam out of your system. In some organizations internal email is subject to review by upper management – so keep it professional and on task. Watch for "copy all" on your replies! Don't use email, or its 21st century alter ego – texting, as a substitute for face to face communications. It's hard to read a person's expressions or body language!

There is no substitute for face time. You can and should use conference calls, net software like "Go to Meeting", and video conferencing to minimize the enormous loss of time and vitality that travel in this age of TSA and jammed delayed flights causes. But you still need to travel some to insure that the voice you hear on a conference call and the face you see on a video monitor is familiar. Travel also gives you a lot of other familiarization time. Normally a site visit involves not only a work day or two but a meal out. When they visit you be a good host. Take them to dinner at a nice place. Involve a few of your subordinates. Let the visitor get to know your group. Get to know theirs.

Security is a concern at work also. And not just for those working in a defense related industry. Remember that there are firms out there that specialize in industrial espionage. These outfits not only try to hack into the computer systems of industry. They send questions out over the internet, try to get your employees to "interview" for non existent but lucrative sounding jobs by pretending to be executive search agencies, tag along on facility tours, show false press credentials, buy their way on site by acting as an employee of a contractor working there, eavesdrop on cell phone conversations, and anything else they can do to obtain proprietary information they can then peddle to your competitors.

Your company's intellectual property is a significant asset. As a manager, you are responsible for maintaining its security.

Your resource base does not stop at the borders of your department. Your company has lots of resources and you must understand how to tap them if you need them. Establishing good communications with your peers is essential. Sharing knowledge, involving them if a situation will touch on their area of responsibility and keeping them in the loop builds trust. You will be amazed at how helpful other managers will be when asked if they feel confident that you will reciprocate when they have a need.

Principle XXVIII: Learn how to speak publicly and how to make interesting and succinct presentations.

Make presentations that inform and are free of technological jargon and unsubstantiated facts. Make them interesting. Use visuals. If you are not comfortable, attend a seminar on making better presentations. You must present to sell your ideas. The trick of a good sales person is to understand their customer's needs and to tailor their pitch to show how their product will meet those needs.

Making a highly technical presentation to a group of marketing people does them a disservice. They care about understandable technology they can market to their customers or something that will free up some more advertizing and promotional dollars – not a gee wiz internal circuit design you may be so proud of.

Presentations to top management are business presentations – not technical side shows. Top management cares about market share, return on sales, competitive advantage, media perception, cost of goods sold, and return on invested capital. Tailor your presentation to your audience. If you do, your ideas will get a lot more favorable treatment.

And your ideas are what sets you apart from your peers and gets you promoted. Public speaking is essential at higher management levels. It is a learned skill also. Many people are uncomfortable speaking in front of a group. They dread it and dodge it anytime they can.

If you really want to move up the management ladder you have to get over it. And the only way to do this is to learn to swim by jumping into water that is over your head. Thankfully there are lots of training courses in public speaking. Usually they function by making each attendee make speeches to the class – over and over on multiple different subjects followed by feedback. Repetition builds confidence.

Have you ever sat in a presentation where the presenter stood behind a podium and read, looking down at his notes and almost never making eye contact with his audience? Fifteen minutes into such a presentation half the audience was taking a power nap.

Have you ever been in a presentation where the presenter stood to the side and actually looked at the screen and read from her slides? Awful technique. People can read a slide ten times faster than the presenter can read one aloud. So 90% of the time the audience was bored to tears.

Get out from behind the podium, face and make eye contact with your audience, and engage them.

What if you get asked to make a presentation with no advanced notice? Tough fix but it does happen. One trick is to engage your audience and get them to help you by setting the stage and then asking them for questions. So they basically are doing your outline for you on the fly and leading you in the direction of those things they are interested in. But if you think there is even the slightest potential that you will get placed into an impromptu presentation situation, have a couple of fleshed out ideas in your mind in advance.

If you have time, take a leadership role with community support groups. Then you get to do some public speaking outside the work environment where the risks may not be so great. You will learn far more than you will help. Yet your help is important.

Principle XXIX: Be consistent and insist that your subordinates solve their problems, with guidance from you, themselves.

Subordinates bring problems to their boss for help. They do not bring them to bosses so that the boss takes them on him or her self. Ok, maybe some do. But do not buy in. Every subordinate problem you take on yourself is one more thing you must deal with.

A very old but effective way to describe this situation is the monkey story. Picture this: one of your people comes to you with a big monkey on his back. You invite the monkey to jump off his back and climb on yours, joining the twenty or so already residing there. Just how much time are you going to be able to spend caring for and feeding the new monkey?

Discuss their monkey with the person. Provide guidance for its care and feeding. With the person develop an action plan for putting the monkey on a strict diet and ultimately eliminating it. Send your subordinate off to deal with his or her own monkey.

Principle XXX: Environmental awareness is far more important than you think it is for any business.

Almost every person in the developed world has some environmental awareness. They are bombarded with media reports on global warming and pollution of the air, water, and land. And they care. Companies that do not have a strict policy related to energy use and pollution elimination will not last long in such a marketplace.

Every thing your business does needs to use the least amount of resources necessary and produce the least amount of waste. Energy cost inflation is inevitable as more and more people compete for less and less resources. As an engineering manager, you have to champion energy conservation and waste elimination. It has to become almost a religion.

If your workforce knows that you will always demand, and will read in detail, an energy balance and a waste generation paragraph in every capital project write-up you sign, it focuses their minds.

Each production site should have already done a detailed annual energy and waste study of their facility. Think of it as placing a glass ball around your property and measuring every thing that enters and leaves for an entire year. Yes, things like raw materials in and finished goods out are there. But so is water in and waste water out; lubricating oil in and used oil out; natural gas in and hot stack gas out; cooking oil for the cafeteria in and used cooking grease out; spare parts in and scrap metal out. The idea is to include everything and develop a balance sheet: in has to equal out. Where did everything taken into your site go?

How many times did you heat raw materials or in-process materials to subsequently cool them expending energy each time? Is that absolutely necessary to the proper conversion of raw materials into finished products? Could you change to a more continuous process and minimize the heating and cooling and reduce in process inventory? How long does it take for one molecule of raw material to go through the process and become part of finished goods? Can this be shortened?

How about environmental air conditioning? Environmental air conditioning is notorious for first cooling air to low temperatures to dehumidify it and then reheating it to make it warm enough to reintroduce to offices or plant areas. Consider other forms of dehumidification that do not use refrigeration – like desiccant dryers.

Insisting on an energy and waste balance is a key attribute of a successful engineering manager.

In summary - use your common sense.

Being a good manager is complex but not rocket science. Establish your own management principles and then follow them.

I will add a few paragraphs about management and the military. It is a common misconception that civilian management techniques and military management techniques are totally different. That is not true. The same principles apply. They may be defined in other terms – in military speak – but they are the same. I invite the reader to look at the following:

The Organization

The organization functioned well. Most of its employees knew their job and were experienced in performing their assigned tasks. Everyone knew how they fit into the organization and how their particular function contributed to the organization's success.

There were clear lines of responsibility and authority. Every department was organized similarly and communication between levels and across departments was clear and concise. Everyone knew the range of salaries for each level of the organization and benefits were excellent, generic and proportional for all levels. There were few private offices and most front line supervisors and managers worked right in the area with their employees. Senior managers did their job: establishing vision, setting priorities, allocating resources, and developing subordinates.

There were well documented and thought out procedures for normal day to day operations and work instructions were clear and precise. There was a clear vision of what organizational success looked like and the vision was communicated and shared at all levels.

Long term and short term plans were prepared with input from all levels and well documented and distributed.

There were measurement tools available that permitted all levels to evaluate their progress towards objectives and they knew how they were being measured against performance to plan. Follow-up was continuous and progress was not only measured but communicated with all levels.

There was a formal system for training new employees. In addition, new employees were always assigned a mentor who showed them how the work was performed until they had a good grasp of their responsibilities.

The organization was flexible and responsive. Since the organization had to perform both mundane tasks and concurrently function in various competitive situations, it was imperative that it could maintain itself yet respond in changing environments.

Permanent task groups were formed from members of multiple departments. Over time and with training these groups became quite proficient at the specific actions required in specific circumstances. Then these task groups could be called up and they hit the ground running to execute the required task.

In the event one group was unable to perform all the tasks assigned other groups automatically stepped in to assist with minimal input from top management. This permitted top management to concentrate on the strategic and tactical decisions necessary for the organization to achieve success and to communicate with other organizations, suppliers and customers for the benefit of all.

Meetings were well organized, had specific objectives, and produced measurable plans of action and commitments from department representatives to do the work assigned to them. There was a simple and understandable financial control system of budgets, including some discretionary funds. The authority to commit resources was understood at all levels.

And although seldom required, there was a documented and graduated disciplinary system to deal with organizational members who were not performing to expectations. Conversely, there was a personal help system to assist organizational members who had personal problems or family problems. Sound like a good organization to work for? It was in a way.

The organization described above was a United States Navy destroyer operating off the coast of Vietnam in 1968: three hundred and sixty four men living and working, sometimes in actual combat with an enemy, on something not much longer than a football field. All that was done in the above description was to demilitarize the different organization and management techniques used to make the ship function well in a changing environment.

Military management is different. The structure and levels of responsibility are more defined. There are few non supervisory levels since the enlisted men and women are also broken down by a rank structure and everyone above Private is expected to both work and lead if called upon. People wear their status on their uniforms with stripes or devices so it isn't difficult to determine who is senior and who is junior.

Directives are more formal and there isn't much leeway if ordered to perform a task. The people even speak another language full of acronyms and idioms that the uninitiated would hear as gibberish.

But many of the management skills that work well in the civilian world work well in the military.

SECTION II

Business Economics for Engineering Managers

Section II, Chapter I

ECONOMIC MEASUREMENTS

There is no need for an engineering manager to be an accountant – except maybe for keeping very accurate track of project budgets, expenditures, and anticipated expenditures. But an engineering manager needs to understand the basics of how business profitability is measured, what the measurements mean, and how they effect the approval of capital projects, development projects and new products.

Sadly, businesses use different measures. The differences are there to make the business look better to potential investors, stock holders, and even owners. However the real true measure of a businesses success is how much true after tax profit is generated per unit of capital employed. This is return on total invested capital (ROIC) and total means equity, assets, and borrowed money – every dollar used to generate the profit. It is a percentage and can be compared across businesses and even with the return on bonds or personal savings accounts.

Many business top managers and accountants have figured out how to "fudge" this by using "Return on Equity" or "Return on Assets". Return on Equity (ROE) is after tax profit per unit of stockholder's equity, so borrowed operating capital isn't in the equation. The result looks good but the business may be leveraged to the nines. Enron looked real good if return on equity was your measurement tool.

Return on Assets (ROA) is another system to measure the after tax profit per unit of assets. So as assets are depreciated, returns go up. One trick here is to sell off an asset near year end, pocket the income and take out the depreciated asset value: fat addition to the numerator and a deduction from the denominator. This can generate a nice boost to the measurement percentage.

There are even businesses that measure Return on Total Assets (ROTA). This is a purest approach where after tax profit is divided by the current (replacement) value of all assets, plants, equipment, inventory, borrowed capital, everything. So depreciation has no effect. ROTA is usually far lower than ROIC, ROA, or ROE.

There are a lot of other tricks and creative accounting methods that I don't pretend to understand. If you sit down and read ten annual reports from businesses you would get ten different pictures based on the accounting methods each used. And they all would look great. Yet some of the businesses may not be as great as they seem.

For an engineering manager you need to know what the driving methods that your business uses to measure itself are, and how those numbers are calculated. Then you will know how your project, product, and developmental proposals will be evaluated.

We will look at some business measurement tools:

Cost of goods manufactured and sold statement formulas:

Prime Cost = Direct Materials Cost + Direct Labor Cost

Where Direct Materials Cost is the cost for all raw and partially finished materials purchased that end up in the finished goods and Direct Labor Cost is the total cost of the workforce that actually manufactures the product.

Total Factory Cost or Manufacturing Cost = Prime Cost + Factory Overhead

Basically Prime Cost plus the cost of the factory's management, facility maintenance, real estate taxes, usually the on site engineering group, on site accountants, security, and all costs associated with keeping the factory ready to produce = Factory Overhead. This also includes the cost of energy, waste water treatment, scrap, etc.

Conversion Cost = Direct Labor Cost + Factory Overhead Cost

Here raw materials are not included but raw material scrap losses are.

Cost of Goods Manufactured (COGM) = Total Factory Cost + Opening Work in Process Inventory - Ending Work in Process Inventory

Number of units manufactured = Units sold + Ending Finished Goods units - Opening finished goods units

Per unit cost of goods manufactured = Cost of goods manufactured / Units manufactured

Materials used or consumed = Opening inventory of materials + Net purchases of materials - Ending inventory of materials

Cost of goods sold (COGS) = Cost of goods manufactured + Cost of opening finished goods inventory – Cost of ending finished goods inventory

Cost of goods sold is an important number and will be different for different industries. The internal effort will be to minimize this number.

Income statement formulas:

Gross profit = Net sales - Cost of goods sold

Per unit gross profit = Gross profit / No. of units sold

Percentage of Gross profit to sales = (Gross profit / Net sales) × 100

Operating profit = Gross profit - Operating expenses

Where Operating (or commercial) expenses are all the selling, marketing, and distribution expenses + general or administrative expenses

General or administrative expenses is where all the otherwise un-captured costs of doing business show up. Things like corporate management salaries and benefits, legal fees, interest on borrowed money, travel, training, corporate office expenses, stock holder meetings, and miscellaneous unidentified costs.

Net profit = Operating Profit – Taxes

It is the profit after taxes that a business gets to take to the bank.

Per unit net profit = Net profit / No. of units sold

Percentage of net profit to sales = (Net profit / Net sales) × 100

Sometimes this is Return on Sales (ROS)

Return on Invested Capital (ROIC) = Net Profit/Total Capital Employed

Return on Invested Capital: Let's face it, ROIC is the reason for business. It is a measure of the business's efficiency in using capital to make real profit.

I'm sure some accountants will take issue with the measurement tools noted above. Each business has its own way and own measurement tools. Usually these have developed over the years to target specific areas the business wants to concentrate on. But any measurement tool has to be targeted to what you really want to accomplish.

The old adage of "you get what you measure" is very true. If your business manages to the quarterly bottom line then it will not be really proficient in making strategic decisions about the long term future. If upper management's bonuses are tied to return on equity you can bet that they will not be averse to borrowing like crazy.

If, like the late 2000's banking industry, top management's bonuses are tied to how well they can package risky investment portfolios and peddle them as valuable derivatives, you can bet that they will not review the risk of the individual investments in the portfolios very well. In fact, individual investment risk will accelerate as brokers try to fill up derivative packages. When these risky investments unravel the whole system will start to collapse.

If your marketing department is not measured with something like net incremental sales per incremental marketing dollar expended, don't expect them to review advertizing expenditures well.

If your feet are not held to the fire for insuring that the dollars expended for you and your engineering staffs have some correlation to the workload you execute against, you should develop your own measurement techniques.

I developed a logarithmic scale that equated engineering staff to millions of capital dollars expended per year. It was adjusted annually for inflation. Small projects always gobbled up more manpower per dollar expended than large projects. Once the tool was refined, I could make a fairly good guess as to the manpower I would need if there was a significant change in the capital to be expended over a five year plan. There were times when I had to reduce staff. I tried to do this by attrition.

Section II, Chapter II

PROJECT JUSTIFICATION

ROI = Net Profit from an Investment/Total Capital Invested to Generate that Net Profit

Projects usually get reviewed based on Return on Investment or ROI. This is the expected total return or profit from the project divided by the total capital cost of the assets necessary to deliver that profit. Sometimes operating capital is included. My opinion is that it should always be included. It is part of the cost. Sometimes this measurement will be called Internal Rate of Return or IRR.

Usually a ROI calculation is done on a discounted cash flow basis (DCFROI) which means that income in out years isn't worth as much today so it is discounted at the company's set cost of money rate. All businesses know this number: it is what it costs them to get new money either in stock dividends or borrowing from a bank. Even if a business is totally internally funded, and few are, there is a cost of money if it is only what they could get by investing it elsewhere.

ROI is used to evaluate the value of multiple projects competing for scarce capital. The higher the ROI, theoretically, the better the investment. Provided, of course, that the ROI is based on real anticipated earnings. Sadly businesses are not very good at being realistic with sales projections. So hurdle rates noted below get inflated to recognize the risk.

Most businesses set up what is called a "hurdle rate". This is the minimum ROI acceptable for a project to be considered. It is usually in the range of twice or three times the determined internal cost of money. My experience says this hurdle rate is in the 20% to 30% range.

This is not a hard and fast rule. If a factory is falling apart, the business may have to replace it to stay in business. Here the return is based on the net profit from continuing to have the old factory's capacity available to make products for sale. The alternative is going out of business or at least partially going out of business. So the hurdle rate here may be 15%. Don't be surprised if, prior to approval of a capacity replacement project, an evaluation of moving the capacity to a lower cost area or off shore is made. Businesses are economic entities and make economic decisions.

Many businesses use multiple hurdle rates. The hurdle rate for an energy saving project or labor cost reduction project may be lower than one for a new product – reflecting the risk inherent with new products. And sometimes there are strategic projects done to position the business for a new opportunity, and these projects may not even require a ROI. But don't kid yourself, somewhere someone in the business has or will evaluate the ROI of a strategic investment even if it doesn't show up in the project proposal.

Some businesses evaluate projects based on the pay back period or turnover rate. Pay back is basically the investment divided by the net incremental average yearly net profit and is expressed in years. Much over

3 to 4 years is not considered a good project unless it has some underlying strategic value.

Turnover is a measurement of how many times per year the investment is recovered. It is the reciprocal of pay back. Rates much lower than 0.25 are not usually good enough.

Businesses have literally hundreds of methods for evaluating capital projects. And capital projects are business proposals. An engineering manager needs to learn how to present them as business proposals.

Spend some quality time with your friendly local accounting manager. Learn what the hot buttons are for your business and how the key measurement numbers are calculated. Run through a *sensitivity analysis* with him or her to understand how tweaking key numbers drives the outcome.

One item of note: In most expansion projects (related to new, line extension or increased capacity for existing products) increasing or decreasing the actual capital (assets, hardware, plants and equipment) cost of the project has little effect on the ROI. The ROI is a lot more sensitive to product price point, volume, and sales and marketing costs. This is most noticeable in the first three years after the product capacity comes on line. Try increasing sales and production volume 10% and watch ROI jump upward significantly. Try decreasing capital cost 10% and notice that the ROI hardly changes.

Many large businesses require different manufacturing sites to compete for capacity additions. Each is asked to prepare a draft project. This system can be the kiss of death for older factories that may need to upgrade older systems and who may have a conversion cost disadvantage based on higher labor, energy, or administrative costs.

And if the older site doesn't get the increased volume, the next time they will be in an even more difficult situation. Sadly, they have become internally uncompetitive. And it may not be their fault. It may be the nature of the area where they are located.

Their internal competition can put a case of product on their back dock cheaper than they can. Unless the older site can sell some attribute specific to their site alone, they are destined to suffer slow but relentless attrition.

Drive around the rust belt and look at how many abandoned factories there are with closed doors, crumbling brickwork, chain link fences, and empty parking lots with grass growing up between the cracks in the asphalt.

Although the city fathers dream that some company will come along, buy the abandoned site, refurbish it and hire 500 local workers, this is not going to happen. Even if they give the company the site and defer or eliminate property taxes for years, it may not be the best economic decision for the business. In many cases it is still cheaper to build a new factory in a low cost area, hire low cost workers, or build the new factory off shore.

There are miracles. But they usually require a whole bunch of incentives like government backed very low interest loans, suspension of closed shop laws, or even outright subsidies.

Jack L. Wells

SECTION III
Engineering Functions

PRODUCT, PROCESS, PROJECT

Section III, Chapter I

PRODUCT, PROCESS, PROJECT

Breaking down the engineering management profession into just these three elements seems like over simplification. And we all know that engineers tend to think in the other direction. However, management is getting things done simply through resource allocation: time, money, and people. Many of the principles apply equally across all aspects of engineering management.

There are differences – differences that cause general management principles to be applied slightly differently. But it is more a matter of emphasis than it is a total paradigm shift. Product engineering requires a high level of creativity at the leading edge. And at the edge creative people need more nurturing than managing. But creativity is also essential in developing processes and executing projects.

To start, let's define Product Engineering as all those elements that encompass the development and preparation for sale of the actual product that the business will sell in the marketplace. It could be a jet liner or a child's toy. In most cases the saleable attributes of the product are not determined by the engineering group – they are determined by the businesses' marketing or sales group.

So the engineering effort involves determining how the product will function and look to meet the marketing and therefore the consumer's need. It also includes the work necessary to support the customer's use of the product, and this in turn includes the quality and service aspects of product support.

Process Engineering is all those elements that involve determining how the actual product will be produced. Some call it Manufacturing Engineering. It is the work done to design the tooling that will make the parts and assemble them into the product. It involves the design of production stations and lines, of special fixtures and assembly tools, and the work flow through the manufacturing facility. It involves all the things done to raw materials and externally sourced items to incorporate them into finished goods. It involves on line quality control and off line quality assurance.

Project Engineering is the application of science to build things for use in the real world. Project Engineering can and does occur in the product engineering arena, the process or manufacturing engineering arena and in support of the research, developmental and production facilities. I define it as any effort that requires the physical modification, expansion, or construction of systems that ultimately produce, distribute, and even sell the companies finished goods or support that effort. Projects have three characteristics: they have a defined scope, they have a defined budget, and they have a defined completion date.

Section III, Chapter II

ENGINEERING ORGANIZATIONS

There are probably as many ways to organize the engineering function inside a business as there are businesses. Usually there is a relatively strict delineation between the product engineering group and the process and project group. In some businesses like pharmaceuticals or food processing there is no product engineering group – that spot is filled by biomedical, biochemical or food science people and is usually a part of the research and development team. Process engineering may be rolled in also.

Quality engineering, safety engineering, and other groups may be stand-alone, or rolled into the product or process groups, or even split between them and project engineering.

The project engineering groups are either centralized, decentralized or some hybrid of the two. The hybrid is normal. I will discuss the advantages and disadvantages of each a little.

Central project engineering is defined as an organization where all significant projects are executed by an engineering group attached to the corporate or division staff. Project engineers are assigned to projects and execute them at the various company sites by traveling there and making it happen. Usually there is a set of engineering standards established by the central group and projects follow those standards. Sites may have a small plant engineering staff that take care of the very small projects, sometimes look after maintenance, and interface with central engineering.

This type of organization is typical for small to medium sized businesses.

Central engineering has the advantage that it can get by with a smaller overall engineering staff. The common standards insure that the plants are similar in technology, and therefore, the business gets some volume advantages in purchasing spares and even raw materials.

Central engineering is staffed to handle the normal corporate or divisional capital project budget. It makes little difference if plant A has a $30MM capital budget this year and $5MM next. It is the total that drives headcount. Naturally, travel costs are high.

Central engineering can also insure that project requests are reviewed and estimated in a similar manner, are in consistent format, and that justification is held to a high standard.

The disadvantage is that a project in plant C done in year five may end up the same as a project done in plant A in year one. There is little incentive for technological improvement. When tasked with adding capacity for plant C, central engineering groups have a tendency to just drag out the drawings for the old plant A project and duplicate it. It keeps the plants uniform but there is not much manufacturing competitive advantage in doing it the same way over and over – usually the same way other competitive businesses in the market have done it at their factories.

There is another disadvantage. The engineers who execute the projects are like sea gulls. They fly in, strut around and squawk a bit, dump an old system on the plant site, and fly out. The site has to deal with any problems in the old technology or in the usually poorly supervised installation themselves. It takes a major disaster for central engineers to be dispatched to help.

The other extreme is totally distributed project engineering. Each site has its own project engineering staff. Usually you will see this in organizations with only one or two sites, seldom more than three, but I knew of one business that did this with twelve sites. Each site executes its own capital projects and staffs its engineering group to handle its average workload.

In distributed organizations the engineers who are executing the projects are on site. They live there and are available to support production, have a vested interest in the success of their individual projects, and are free to explore new technology. In fact, the engineers compete with engineers at other sites to deliver new, better, improved, less costly, and innovative technology systems. Many of these innovations can have a significant effect on the site's manufacturing cost.

The disadvantages are that a distributed group usually has a higher headcount when taken in aggregate. Some companies live with this since they consider the distributed engineering group as a good source for future production managers and have made a strategic decision to get more technically qualified people into production management. You will see this happen in highly automated factories.

Distributed groups build different plants. So different that each plant becomes unique. Transferred production, quality and safety managers have a significant learning curve as they try to cope with the idiosyncrasies of the actual hardware and software at another site.

Leveraging one site's innovations by installing the same technology at another is very difficult. The operators and technicians need retrained, the stockroom doesn't support the other site's equipment, and the engineering group develops a "not invented here" syndrome. Seldom is a transferred idea as successful as it was at its originating site.

Finally, the same product made at two or more different sites becomes different since it is being made on different equipment and maybe even with different processes. Maintaining consistent product quality across the business is very difficult.

Consequently, most businesses end up with a hybrid organization. They have a central engineering group and they have small distributed engineering groups at the plant sites reporting jointly to central and the local site's top management. Large projects are executed by central with help from the site's engineers. Small projects are left to the site to do.

Central sets standards, evaluates project proposals, coordinates technology development, and recommends training. Usually central coordinates performance evaluation. If it is a very large business, there may well be divisional central engineering groups and an overall corporate group. All corporate does is set standards and help with major project development and evaluation. Corporate is usually tasked with advising top management on technical matters and insuring that the technological vision established by top management gets disseminated through the organization.

Hybrid engineering seems to function the best, eliminating most disadvantages of the extremes. No organization is perfect. Your business has, hopefully, developed its own which leverages the strengths necessary to make your business successful.

One final item on organizations: most engineers who transfer from one industry type to another are amazed at the similarity of the technology employed to produce widely divergent products. A continuous process industry may be different from a discrete component industry in many respects but there are still a lot of similar equipment, systems, and software.

Equipment vendors have long ago learned how to modify their designs to meet the needs of disparate industries. Don't think your business is so unique that it requires an engineering staff of totally industry specific engineering professionals to function. A good steel industry engineer can quickly transition to a food processing industry. He or she needs to learn the different sanitation standards and processing techniques. But given an experienced mentor, he or she can learn to function quickly.

Bringing engineering talent in from other industries has another advantage. It inserts a different perspective into your group. The new engineer may have some new ideas you can exploit. The wider the experience base of your team the better.

Section III, Chapter III

PRODUCT ENGINEERING

Product engineering is a continuous set of economic tradeoffs. The most technologically advanced gee wiz product is worthless if it costs more to develop and produce than the marketplace is willing to pay. And the marketplace is very good at determining value for money. Give it a quality product that meets a customer's need for less than the competition and the customer will buy it. Give it a product that is low quality or too expensive for the need it meets and it will languish. So a product engineering group is always inside this trash compactor watching the walls converge.

Built into the economic equation is the cost of manufacture. A well designed product is one that can be produced for a reasonable cost in light of what the customer will pay. So the trade off between raw material cost and manufacturing cost is always there. If your product is gold jewelry your raw material cost will be high and manufacturing cost low. If you are making napkins it will be the reverse. But each has to minimize the cost of goods sold.

Cost of goods sold is the combination of the cost of raw material, manufacturing including not only labor but the amortization of the capital expended to provide the production facility, the energy expended, the waste generated, and the production site overhead.

To this may be added any distribution cost, advertizing, promotion, corporate overhead and the amortization of the cost of the research and development that went into developing the product.

For simplicity let's divide products up into three types. New products, line extensions, and existing products.

New products: To a certain extent this is where the fun and risk is. Whether the concept comes from a market survey, a customer request, or is an internally generated entrepreneurial idea; you basically get to start with a clean sheet of paper. You get to adapt existing technology or develop new technology. You get to hopefully get a totally new product into the marketplace with your name on it.

Very few totally new products are successful. And if they are, the marketplace quickly generates lower priced clones and your initial 100% market share erodes. Over time the number of competitors fall off as prices fall and margins shrink. Hopefully you remain – but in so many cases the original developer of a totally new product gets squeezed out.

There are multiple examples historically. The automobile would be one. Ford, General Motors and Chrysler may still be around (barely) but what happened to LaSalle, Packard, Studebaker, Stutz, Hudson, Oldsmobile, Plymouth, and now Pontiac, Saturn, etc.? Do you remember the software spreadsheet VisiCalc (Software Arts)? It was the first PC based spreadsheet in 1979. How about Lotus 123, Multiplan, Quattro Pro, or SuperCalc? You can probably come up with many more examples from high volume low cost consumer products like a toaster to low volume high cost military aircraft.

Line extension "new" products are far easier and cheaper to develop. Adding a canister vacuum cleaner to a line of uprights would be an example. Coming out with a station wagon version of a coupe. Adding an automatic coffee grinder to a current drip coffee pot. Putting an integral scanner into an existing ink jet printer. Line extensions leverage the existing brand's name and recognition and require a lot less advertizing and promotion than new products. Hopefully they require less manufacturing assets also.

Existing products only, hopefully, require little development. The need is for more production capacity although any new plants may require small changes to the products themselves to improve and lower the cost of manufacturing.

Then there is the ongoing effort to keep products current. Every modification does three things: eliminate bugs in the previous product design, add some bells and whistles that marketing can sell or that improve the product's cost structure, and add all new bugs. Sorry. That is what usually happens.

Most organizations are charter members of the "let's turn every bolt another quarter turn" club. Removing one penny per unit from the cost of goods sold adds one penny to the bottom line. If your business sells 100 million units, that's a million dollars. A significant incentive. Yet there should be a fear that the one penny cost removal will do something bad to the product. No one wants to kill the goose that lays the golden eggs.

This is where "triangle tests" come in. They are used to see if there is any perceived difference between the existing product and the cheaper to manufacture product. The raw material could be cheaper to obtain or the actual manufacturing method could be cheaper to execute.

The original product and the new product are provided to consumers and they are asked to describe the differences they see. In most cases they will see no difference. Surprise. That's what you wanted. So the new "improved" product becomes the standard. A few months later another cost savings idea is proposed and goes through the triangle test successfully again.

This entire effort causes something called incremental degradation. The founder of the most successful private candy company in the world told me once that triangle tests were a way to kill a successful product over time. He said that using triangle tests he could change a table into a chair over seven years and no one would notice. He was being a bit facetious but a string of individually imperceptible changes to a successful product can make it so different that it ultimately degrades and no longer delivers the value for money that made it successful initially.

An example: The strong, well built, heavy, long lasting, good fit and finish automobiles produced by the US auto industry in the 1950's and early 1960's. Many of these cars are still on the road in Cuba. Some professionally restored ones sell for millions at the Mecum Auto Auctions. Then compare them with the tinny, unreliable, poor fit and finish vehicles produced by the same industry in the late 1990's. And consumers noticed. Always searching for the most value for the money expended they shifted to foreign made vehicles.

There are multiple reasons for the decline of the US auto industry, but incremental degradation of the product is one major one. If they wish to survive, it is time to improve the quality of the products and the value for money delivered. And some are starting down this path while dragging a legacy of expensive and excessive pension and benefit entitlements behind them like an anchor. Will they ever recover the market share they lost? A bit – maybe. All - never. The consumer has a very long memory and the entitlements reduce the value for money they can deliver and still be profitable.

If their current vehicle is delivering good reliability and value for money consumers have little incentive to risk being dissatisfied again. And all the "buy American" ads on TV will not perceptively change this. Remember, most Asian and European automobile manufacturers manufacture or at least assemble their vehicles right here in the USA now. It is becoming cheaper in light of escalating labor and tax rates in their home countries and upwardly spiraling shipping costs.

No business is immune to allowing itself to ignore a potential problem with one or more of its products. Market leaders have lost their position through tiny product malfunctions that showed up due to the driving desire to save a penny. Toyota may well have hurt itself beyond full recovery with the 2009-2010 surging and sticking gas pedal problem. And it just got worse. More models were involved; the initial stonewalling by top management got them a lot of bad press, and even their top Lexus SUV was listed as a roll over risk. One unresolved problem quickly leads to more negative attention and investigation and bad press coverage on every thing.

Audi went through this in the 1980's with a cheap low quality Audi Fox followed by an Audi 5000 luxury vehicle that allowed shifting into drive or reverse with out a foot on the brake and "ran away." Toyota would have done well to have learned from Audi's mistakes.

A company can get into a fix with a new product that its targeted customers don't know how to use. In 1960 Chevrolet came out with the rear engine Corvair. It was immediate hit for anyone who had to travel snow covered roads extensively since its traction was superb. But Americans had been driving understeering, front engine, rear wheel drive cars forever. Unless they were previous Porsche owners, they didn't understand how to drive a rear engine rear drive vehicle, especially at the edge of tire adhesion. When they went into a curve too fast and got nervous when the car started to oversteer, they hit the brake. The car squatted nose down and the heavier rear end passed the front end instantly, usually followed by an off road excursion. Porsche drivers knew that more power was the right answer in this situation and to never brake in a curve – brake before the curve and power through it. Accidents ensued, Ralph Nader published "Unsafe at any Speed," and the Corvair was dead. Yet in its Corsa version, especially by 1963, the Corvair really was an American Porsche. If it had been launched as a specialty "sports car", not as competitor for Ford's conventional compact Falcon for grandma to drive, it might have survived.

Consumers are very, very good at discerning quality and value for money. Even if your product is something that goes to another business for incorporation into their product, they are your consumers and will notice. If a change you make causes them manufacturing or reliability grief you can bet that they will call you to task for it, if they don't just stop purchasing from you in retaliation.

Another thing to watch carefully and interpret properly is consumer focus group results. Like any other questionnaire, there is the fact that the way you ask a question determines the way it gets answered. Just as who you ask determines the way it gets answered. If you ask teenaged girls what they want in a candy bar they will say they want one that tastes just like a Snickers and has zero calories. Surprise. Of course those two characteristics are mutually exclusive.

Every consumer group has different wants and needs in a product. You have to either focus on one group or attempt to make the tradeoffs that will broaden your consumer base. I would strongly recommend that every manager involved in managing the product engineering function spend some time observing consumer focus groups in action.

Design for manufacturing is a concept that if done well can deliver a significant reduction in the cost of goods sold; but never at the cost of product quality and functionality. As a business you do two things. You make products and you sell products. You have to do both very well. But you can not remove product attributes that the consumer desires just to make the product cheaper to manufacture.

Walking a mile in the other's moccasins is an old Native American saying. The best product engineering managers are those that have done a stint in process or project engineering. It helps them understand the trade offs and gives them creditability. Then when they meet with the other engineering people, which is a must during the product design stage, they can influence the product design effort in a way that maintains product integrity yet minimizes manufacturing cost.

Design for serviceability. If your product is going to require routine maintenance, insure that it is easy to perform. This sounds like a no brainer but have you ever tried to replace a chain saw chain? Some you can do with the removal of one jamb nut – correct sized wrench provided in the package. Some you have to disassemble multiple guards, take out sheet metal screws, and demonstrate considerable mechanical skills. Why? Unless your customer base is only experienced mechanics, you may well find that when the customer is ready to replace his chain saw he will look at other brands.

You can get away with a poor design for serviceability if you have a niche product. The Sunbeam Tiger, an adaption of the British Sunbeam Alpine comes to mind. To appeal to the muscle car orientated US market in the 1960's Sunbeam shoehorned a 260 cubic inch Ford small V-8 engine into a small two seater sports car originally designed for a little four cylinder engine. It went like a rocket.

Yet to change out the spark plugs one had to pull the engine. The last two spark plugs were inaccessible behind the cut out firewall. A pain, but then the torque of the V-8 was more than the clutch assembly could cope with for long so most owners replaced the spark plugs when the engine was out for routine clutch replacement. Sunbeam is gone (early 1980s) but it was for a lot of reasons, not necessarily attributable to the Tiger. And the Tiger was a low volume, high priced, niche product that did make Sunbeam some money at the time.

Managing people in the new product development arena requires a unique blend of coaching and encouragement. The manager has to accept the fact that there will be more failures than successes, that prototypes will fail and that costs are difficult to control.

Innovation requires creative people stressed just enough to stimulate their creativity but not sufficient to stifle it. It is a fine line – like walking barefoot on the top of a picket fence. It can be done, but oh so carefully.

If you are managing a basic research group in a government or industry association funded location, you are more involved in writing funding grants and evaluating publication potential than in stressing your employees to produce. But in a business, where the ultimate goal is profits from new products, you have to keep the workforce focused on time and money. So sometimes you have to be the person with the whip – applied precisely and with just the right force behind it – but applied none the less.

Any new hardware product has to be a unique blend of mechanical equipment and electronic control. Using electronics to fix an inherent mechanical problem or the other way around is a mistake since it usually makes the product more complicated than it needs to be, or unreliable.

Every electrical engineer on your staff is, or at least should be, capable of using a computer and servo to build a highly accurate variable profile cam shaft. And every mechanical engineer should be capable of designing a mechanical linkage that can be made to function as a variable profile cam shaft. But neither may be correct. Use as much sophistication as is essential, but no more.

Minimize adjustments in the final product design. It is understood that there will be lots of adjustments in the prototypes but engineer them out in the final design. If there are adjustments, your consumers will adjust them. In many cases a misadjusted product will not perform properly and will be considered unreliable by the consumer. Make the product do its own adjustments for wear. Make it easy to operate and maintain.

Team work in new product design is a given. Teams of people from various disciplines should meet often to share ideas. These tend to become brain storming sessions if facilitated well, and the group will push the outer edge of the envelope. During the initial concept development this is essential. When you get to the prototype stage, the same group becomes an evaluation and problem solving group.

All new products target an opportunity window. If your marketing types see an opening it is just a matter of time before your competition's marketing types see it also. The company that gets their product to market and at least into regional and preferably national distribution first gets all the good press and initial impulse buyers. The company that comes in second gets to spend a lot more advertizing and promotional dollars to try to pry a bit of market share away from you. Or they have to price their product a lot lower than yours to motivate buyers. So quick and precise is essential. This means new products are very expensive to develop and launch, and it is why top management is very careful when selecting ideas and giving them the green light.

Some other rules of thumb for new consumer products: Your company will spend two to ten times your gross sales of the product in year one in advertizing, promotion, slotting allowance and distribution to go national. Year two will be near twice gross sales. Year three is break even. Year four you make a little. Year five and onward you make a lot. And in the consumer product world, one out of five new products survive beyond year five. They die for multiple reasons – usually because they are starved to death for lack of advertizing and promotional funds. It is a huge financial risk. But if you get a good one it can be an enormous long term reward. Better odds than playing the lottery – but not a lot better.

New products sold to other businesses for incorporation into their new products are similar but your company is not ponying up the launch dollars. Expect your customer to be a real hard nose on the cost of your product. They need to be.

Reconnect with the marketing and manufacturing group often. Any development team has a tendency to keep adding complexity and unneeded functionality to a new product that may not, after the additions, be saleable, or salable at the target price.

And you sure don't want to design a super product that has to be manufactured by hand unless you are planning an item to compete with Rolls Royce automobiles and can sell the product for $500,000 each.

In almost every case your product and the process you use to manufacture it will have to be certified by the International Standards Organization (ISO). It is a very good way for purchasers, either consumers or intermediate manufacturing users, to insure that what they are buying from you has been designed and built to international safety standards and that the quality they want is built in. There may be multiple other standards you must deal with. Don't scrimp on these efforts. Your company's reputation is at stake. One screw-up and you will be doing your company and yourself a significant disfavor.

In summary, a new product development group is tasked with developing a unique product to meet consumer needs under strict time constraints. The product needs to hit the market in precision form. It is expensive and risky. But successful new products provide significant profit rewards.

Section III, Chapter IV

PROCESS ENGINEERING

The design of the actual equipment that fabricates the product you are going to sell is a critical function. If you can design a reliable system that can produce your product efficiently and with high quality built in, you achieve a manufacturing competitive advantage. This advantage allows your company to produce at a lower cost than your competitors and can give your business more working capital to use for advertizing, promotion, or even salaries to keep your employees happy.

Continuous process industries usually have a group that determines, working with the product design people, the required alterations to raw materials essential to make the product. The key is "essential". Processes should be simplified as much as possible. If you have to heat the raw materials, see if you can heat them together after mixing, not separately. Once you heat them and if you have to cool them, can you recover the heat by using already heated in process materials to preheat those entering the heating process. If somewhere you are exhausting steam, why not recover the heat from the steam prior to exhausting it?

The more different raw materials there are in your product the higher the manufacturing cost. A good process engineer will challenge the product design people to evaluate the functionality of each raw material. Is it really there for a reason that measurably contributes to the final functionality of the product or is it just something they would like to put on the ingredient list? If it is an exotic ingredient, is it essential in that form or would a generic work just as well? Remember that ingredient suppliers send thousands of raw material samples to the development groups of continuous process industries. They don't send their cheapest ingredients unless forced. They send their most profitable exotic ingredients.

If your process inherently makes some rework, and you can't engineer this out of the process, then how and where should you reincorporate rework into the primary process stream? If it makes scrap then how do you convert scrap to rework? It should be noted that most rework and scrap come from system start-up and shut down. It is the transient periods that cause the grief. Process systems all warmed up and operating at steady state don't make much rework or scrap. Systems need to be designed for controlled, linear, start-ups and shut downs.

Continuous processes are always more efficient than batch processes. However, don't get fooled into designing ultra complex machinery that jams a bunch of discreet fabrication elements into one long machine with no surge capacity between individual major processing steps. If you have a process element that does one key thing to a product, provide some surge capacity at its discharge or the next step's input.

Consider this: If you have a processing element that has an inherent 95% efficiency factor and you close couple it to nine other processing elements that each have a 95% efficiency factor, then you have an overall machine with a 60% efficiency factor: $0.95 \times 0.95 \ldots$ or $0.95^{\wedge}10$. I have seen this attempted many times. The low overall efficiency of the machine doomed it to failure each time. The machine also made enormous quantities of scrap.

Rotary motion is almost always preferable to back and forth motion. Tolerances are less critical, transients are better controlled, and speeds are usually higher. If you couple rotary motion with computer servo control, you can construct a very efficient high speed rotary machine.

PRODUCTION LINE DESIGN

Each industry, and even company, has some guidelines for the design of its production facilities. The most restrictive of these are for the design of the actual fixtures, pipelines, mixers, tanks, or fabrication stations within the facility. If you are working in a pharmaceutical or food, both human and animal, industry, you have to design your process for periodic cleaning and sanitation. These industries try to do as much cleaning as they can using Clean in Place (CIP) where mixers, tanks, pipe and etc are emptied of material in process and then washed internally with cleaning solution and water automatically, or at least semi automatically. CIP design requires care because the incorporation of even a small amount of cleaning chemical in finished product is a product contamination issue. Everything has to be designed to drain completely. Usually they then need to be flushed thoroughly and dried completely with clean compressed air.

Pharmaceutical and food factories must be designed for total clean ability. This means minimizing everything mounted on floors and walls. It means single pedestal mounts for pumps and process equipment, preferably with electrical power and other utilities enclosed in the pedestal and coming up through the floor. Every junction of floors to equipment or walls has to have a minimum ½ inch cove radius. Any area that can catch and retain dirt or product residue will catch it and be a site for the growth of bacteria.

Microbiologically, most bacteria have to have three things to survive and grow: moisture, food, and the proper temperature. Industries who's raw and in process materials are subject to microbiological contamination have to be vigilant. Usually they attempt to limit moisture and control temperature. Even so, their raw materials can arrive already contaminated and there needs to be processing steps to kill any and all microbiological contamination before it gets into the process stream.

Factories that have lots of dust in their process must be designed to safely deal with the dust. Most dust, if enclosed and mixed with the proper volume of air, will explode with devastating effect. Coal dust, powdered sugar, grain dust, or for that matter any material in powdered form that will bond with oxygen in an exothermic reaction is subject to dust explosions.

Dust collection bag houses are very susceptible to explosions. They need to be designed very carefully and the system feeding them has to have quick reacting back pressure and vent valves and overpressure shutdowns. But the real key is to design the systems to keep from achieving the critical air/dust mixture necessary for an explosion to occur and design to totally eliminate any chance for an initiating static spark.

Few companies can dedicate one line to one product only. The lines need to be flexible. They need to be able to fabricate a soda machine today and a coffee machine tomorrow: make cookies today and cupcakes tomorrow, etc. This necessitates the dreaded change over. Changing a line from one product to another usually goes quickly. It is the start-up after a changeover that causes the grief. And start-up grief can be mostly attributed to adjustments.

So many lines are made to be flexible by the addition of slotted hole adjustments. Adjustments to conveyor rails, fabrication equipment input and output tooling, almost everything. And adjustments need to be made during product flow. This takes time and many industries are used to a low slope start-up curve after a change over.

Consider making change parts and eliminate slotted holes. Remove fixture A, install fixture B and start making product B. Attach fixtures to the frames with hex socket cap screws using the same mounting holes. Store fixtures in a common location and do preventative maintenance on them while they are in reserve and idle.

Any adjustments absolutely required need to be equipped with attached metal rulers or color coded holes for tight fitting pins. Again, we want to eliminate, or at least minimize, adjustments necessary at start-up.

DESIGN for RELIABILITY

Production lines, fixtures, and change parts should be designed to run without breakdowns. Breakdowns cost a fortune. When they happen your production employees stand around and watch maintenance untangle all the collateral damage, trouble shoot, retrieve parts, repair, and restart the equipment. But the real insidious cost is that unreliable production equipment is lost capacity and lost capacity has to be compensated for with excess capacity somewhere else. Or the line has to run extra, sometimes at overtime rates, to make up what didn't get produced on standard time.

Preventative maintenance is a good tool. But usually this means that you have a lot of mechanics on back shifts checking idle equipment over and over. The right answer is predictive maintenance. Figure out the life cycle of equipment and replace bearings and sprockets, etc. just prior to failure. Initially design equipment for long life cycles with automatic lubrication systems, robust parts, totally debugged software, and performance instrumentation.

Performance instrumentation should have a "first out" capability to insure that the first excursion from normal operating ranges gets logged first. In most failures there are multiple alarm annunciations and you need to know the first to do any quick trouble shooting.

Personally I have an aversion to Philips head screws and hex head bolts. The Phillips heads invariably get chewed into a cone that needs drilled out and replaced (naturally at the worst time when the line is down for repair and an entire production staff is standing around while the meter runs) and the hex head bolts get chewed into a circle by tools like channel locks and vice grips. I banned channel locks in individual tool boxes at factories where I was responsible. Channel locks were available if needed but kept in the tool crib and painted international orange.

If your industry uses stainless steel screws and bolts this happens even faster. Stainless steel doesn't stand up to high strength tool steel screw drivers and wrenches very well. Hex socket cap screws, especially those counter sunk flush, seem to stand up to multiple repetitive tightening and loosening far better.

Beam clamps were also painted orange and kept in the tool crib. Then when they were used for a quick fix they were visible and could and would be removed and a permanent fix executed.

Fixtures and tooling should be fabricated from the appropriate material and welded into solid units. Bolts and screws should be strictly for mounting: shop connections welded, field connections bolted.

I have a total aversion to "V" belts on any production equipment. They stretch, slip, case harden, break, and require attention and replacement. I feel the same way about roller chain. High strength timing belts with tension idlers are better unless you have a high load application that requires roller chain. Even then, double or triple chain and hardened sprockets and idler sprockets for take up.

Many production facilities rely on hydraulics. Only one drawback: all hydraulic systems leak oil. Expect it. Atomized hydraulic oil mixed with the proper volume of air is explosive.

Many factories use compressed air for everything. Take a walk through your factory during a shutdown. How many places do you hear compressed air escaping? Down in the power plant there is a big air compressor chugging along using energy to supply this escaping air. Is it necessary?

Software control is critical for any system. There should be one official version not only in the equipment's operating memory but available for download on the facility's network. How many problems with your personal computer can be solved with a reboot? In many instances a technician will use a programmable logic controller (PLC) or supervisory computer to troubleshoot a malfunctioning process system. They force contacts or add memory registers to see what happens. That's great. But you still need to go back to the original approved program once the troubleshooting is complete.

Programs should be devoid of all the extra hooks, loops and registers that get built into them during the development process. Once debugged the program should be cleaned up so that it is more compact and understandable. A few simple hooks here and there for troubleshooting are Ok, but a program in development is normally twice the size of what is really necessary to control a system. If you are using adaptive control loops as more and more industries are, these are quite difficult to troubleshoot. There should be historical registers to show where the program was and where and when it modified itself. A malfunctioning sensor can send an adaptive loop's output out of range in a hurry. Garbage in equals garbage out.

Robotics are great. They don't take breaks, vacation, or need benefits. But if not maintained they will make the same mistake forever. Most are very reliable. It is your special fixtures at the end of the robot arms that contact the product, get the wear, and have to be perfect.

Overhead conveyors are very good. They take advantage of the cubic space in your factory and can allow for odd shaped line flow design. They also need maintained. And if they need maintained, or if they can jam and need jams cleared, they need to be accessible for maintenance. This involves catwalks and ladders.

Discrete parts bins are super. A bin designed to securely hold only one part or prefab assembly stops errant parts from getting into the wrong system and can be locked in place for robotics to pick the parts up one at a time.

Work flow on a production line should be as continuous as you can make it. But continuous doesn't mean no surge capacity. You need strategically placed surge so that one little blip doesn't shutdown your entire line. My rule of thumb was a minimum of three times the per minute throughput rate of surge between each major element of a process.

Design for the workers. You wouldn't want to lift 50 lbs from the floor to over your head for 8 hours every day. If you have a work station that requires that, provide equipment to help the workers. It's called ergo metric design and there are lots of books and online information on this.

Safety, safety, safety. This is a no brainer. Make your factories safe for the operators, cleaners and maintenance personnel. You would be amazed how much an operator wants their line and the particular piece of equipment they are operating to run and keep running. And you would be amazed at what they will do to keep it running.

If their hand will fit into the equipment they will put it in there to clear a jam or adjust something. Make it fool proof. Guards need to be interlocked. Open one and the machine shuts down. And if there is a piece of equipment that is unreliable, watch for pennies taped to prox switches, strings holding micro switch arms in place and other such safety guard overrides. They are accidents waiting to happen.

Most factories preclude the wearing of jewelry, rings, watches, ties, or long hair in pony tails from the factory floor. They can catch in equipment and produce some awful injuries or even death.

Every factory needs an effective lock out tag out procedure. This means that the lock out device must be right next to, and carefully identified with, the equipment that it serves. Locking out equipment at a circuit breaker panel only is just asking for the wrong circuit breaker to be turned off and secured while the right breaker is ignored: another accident waiting to happen. If you ever must run equipment during cleaning or trouble shooting you must have a safety observer watching and standing immediately near an emergency stop button. Never rely on the lock out device near the equipment for an emergency shutdown. These "safety switches" are usually not built to interrupt flowing electrical current and may explode if switched under load.

Every motor driven system with integral brakes, especially "power to open spring to close" brakes, must have a way to easily release the brake without power. If an employee is caught in such a system you must have a way to reverse the equipment to extract them.

All equipment must be designed with safety in mind for not only normal operation, and start up and shut down, but also for cleaning, trouble shooting, and maintenance.

Factories are noisy. Some are very noisy. The government wants you to engineer the noise out. Only then will they allow you provide personal protective equipment to reduce the effects of sound overpressure on the workers.

Sound is usually caused by the friction between one item and another or the impact of one item on another. Where possible, limit this or provide sound absorbing control. Low frequency sound is far worse than high frequency. Deep bass will penetrate an eight inch concrete block wall. High treble can be stopped by a piece of paper. And sound shields need to be tight. A one inch hole in a 30 dB sound shield makes it a 10 dB sound shield.

Material selection is important. And stainless steel isn't stainless. In contact with anything containing chlorides, especially high temperature chlorides, it will corrode. If it is under stress you will get stress corrosion cracking. Steam contains chlorides, depending on the boiler water treatment chemicals you use. I once was involved with a factory that handled hot salted tomato juice enroute to the filing line. Stainless steel piping turned into perforated pipe in six weeks. Ultimately we had to use titanium.

Surface painting in a corrosive environment is hit or miss. Yes, some of the more exotic epoxies hold up. Baked on powder coating works well. Most others will fail quickly. Aluminum will grow aluminum fuzz in many environments, including sea water or salt spray. Galvanized steel may be an answer but one tiny nick, field added bolt hole, or touch with a welding rod starts progressive rusting under the coating. In some industries smooth carbon steel with a continuously maintained light oil covering is the best bet.

Every industry is different and yours has developed its own standards for minimizing corrosion and the seemingly endless necessity for replacement or repainting.

Section III, Chapter V

PROJECT ENGINEERING

Most colleges and universities graduate engineers who are technically competent. They know the theory of their fields and can apply mathematics to both describe and apply those theories. But few universities teach Project Management as a course. Engineers have to pick this up through in-house training, the school of hard knocks, and through osmosis. And it is a learned skill. Once learned and applied consistently and with persistence and attention to detail it will serve them well.

An engineer carries a significant amount of responsibility in any business. He or she is the chief technology interpreter for the business. He/she must be able to explain it in laymen's terms to the rest of the business. He or she is expected to be up to date with both old and new technology and be capable of applying it quickly at a low cost to meet business needs. He or she is expected to be able to direct the repair of damaged equipment, look after or advise on the physical plant structure, equipment and systems, and understand and apply codes, standards, laws, and regulations to assist the business.

He or she is also expected to understand the business, not just technology, so that he or she can ask the right questions to enable technology to be applied properly. One of the key areas an engineer is responsible for is Project Engineering Management - turning ideas into reality to meet business needs.

The techniques utilized to be an effective project manager are not all that different than the techniques utilized to be an effective anything manager. However, project management, done correctly, is not an easy task. The project manager must, of course, be technically competent - not only in their own field, but to at least some degree, across the full spectrum of engineering disciplines. The project manager must also be well organized, understand where to go for help, and be capable of getting things done through other people.

Effective project management is so critical to the success of a business that we will devote a significant amount of the entire book to this one item.

The five stages of a project:

Enthusiasm

Panic

Search for the Guilty

Punishment of the Innocent

Praise and Honors for the Non Participants.

Anonymous

The above is not true - although it seems that way sometimes. It is true that most engineers are expected to engineer things. Yes, they are also expected to be technically competent, innovative, creative, and all those other attributes nearly impossible to concurrently maintain.

But they are really expected to **_MANAGE_** projects. An engineer is expected to take concepts and business desires and turn them into operational systems to meet business needs. To do this they are given resources, and they are expected to carefully manage those resources. Deliver - on time - on budget.

An engineer also needs to know that a mismanaged project may seem to have a lot of blast effect at ground zero where he/she stands, but it also has a lot of collateral damage and wide spread fallout. Significant overruns create grief for the entire business - not just the project team and business unit. Missing a time schedule ripples through the organization like a tsunami. Every group involved suffers increased costs and credibility losses. Future opportunities are adversely affected. Not pretty.

The goal of this chapter is to provide a blueprint for effective project management. To start with, let's take a look at two equal projects ...

A Tale of Two Projects

Once upon a time, a long time ago, in a galaxy far, far away ... there were two moderate sized capital projects developed and approved for two unnamed consumer product production sites in the same company. Although initial plant cost estimates were slightly higher, each plant agreed that they could do the required work for $500K. Each project involved the modification of an equivalent production line for a line extension product. An individual project manager (PM) managed each. For the sake of simplicity we will assume that each project manager resided at his or her plant site (distributed engineering).

The project managers had roughly the same number of years of experience since college. At each site both projects were staffed with a mechanical engineer who was assigned as the project manager, an assisting part time electrical engineer and a part time packaging engineer. Each had an operations manager part time also. Each had a couple of "job shopper" process designers and a part time packaging line designer. Naturally, the project teams also had other projects and administrative duties - but the workloads were equal.

Although the projects were somewhat complex, i.e.: add some new raw material handling items, process items; provide change parts to forming equipment; and provide change parts to the packaging system; there were no obvious technical challenges. Both projects were approved at the same time, and start-up was scheduled for 10 months later to meet the rollout plan. Since the plants were in the same geographic area, contract labor, construction material and shipping costs were roughly equal. We will call them Project Alfa and Project Bravo.

Project Alfa got started well. The project scope was defined. The goal: Get it running with minimum start-up costs and make product for distribution to coincide with the marketing launch plan. The budget seemed Ok with a little slack. Timing was a bit tight, but doable.

After getting all the product configuration data from research and packaging data from marketing, the PM set up a budget, met with his team to explain expectations, and got started. Operations wanted a new floor under part of the process since it was deteriorated ($30K). The designers assigned to the project felt like they needed to upgrade their Computer Aided Design software to speed the piping design. Since there appeared to be sufficient funds for these items, the PM agreed, and the software was purchased ($5K). The floor would be done in two months over a scheduled weekend maintenance shutdown.

Process equipment was specified, and ordered. Process piping design started. Electrical design awaited process and packaging design completion. Packaging change parts were specified and ordered with an 8-month lead-time. Things appeared under control and the PM reported so to his manager in his monthly report. The second month's accounting sheets showed $250K allocated, $250K unallocated.

The new floor went in as scheduled but the invoice included an extra $20K for a total of $50K due to an operations supervisor who was on site over the shutdown asking the contractor to extend the new floor into a storage area. The PM was not on site during the floor installation, and was very unhappy but purchasing told him that if a company representative verbally told the contractor to do something, it was a verbal contract and the bill needed paid. That put a dent in the budget. His superior chastised the supervisor - but the damage could not be undone.

Marketing then asked for the carton configuration to be changed from 24 count to 12 count. Sales had advised that 12 count would move faster and appear like product was turning over on the stores' shelves faster. This doubled the number of cartons to be handled per unit time and required a redesign of change parts by the vendor.

With time tight, the packaging engineer called the vendor and said do it. It also required a carton conveyor change to handle the additional volume. The packaging engineer mentioned this to the PM, but did not estimate the additional cost. The monthly engineering report indicated that the budget would be tight, but everything was on time. End of month 6 accounting report: $290K allocated, $210K unallocated.

Then things started to deteriorate. Even though the process design was technically excellent, process-piping design was behind because of bugs in the new software. When finally issued for bid, process piping came in 8% higher than estimated and the most reliable bidding contractor indicated that he was very busy and really did not want to do this job - although he would if pressed using labor from another office.

A vendor for one of the key items of process equipment advised that there might be a delay in shipment due to a potential strike at one of his supplier's factories. Raw material samples were late and when they arrived the viscosity of one was twice the original - necessitating a pump capacity change. Electrical advised that they were way behind because of the process delays and that the new pump needed all new wiring to meet code. The operations supervisor assigned to the project was transferred to another site and a new person was assigned.

The PM working with Purchasing went after the process-piping contractor - cajoling him to reduce his bid. Ultimately some items were removed from the contract, less overtime was anticipated and the bid was almost back in line with estimates. Electrical bids were 10% higher than estimated

The electrical contractor refused to reduce his lump sum price but agreed to do the work on a "time and material" basis. He anticipated that he could bring it in close to the original estimate if everything went well but did not want to take the financial risk of a lump sum contract. The electrical engineer agreed. The monthly engineering report indicated everything was on schedule but the budget was still tight. End of month 9 accounting report: $430K allocated, $70K unallocated.

A short 4-day holiday shutdown was scheduled for installation. Later this would reduce to 3 and a half days due to high production demand. Installation work commenced and the critical process equipment arrived just the day before. The cartoning change parts did not fit - it turned out that the maintenance people had modified the machine to improve reliability a year ago, and it was not on any print. Lots of cutting and fitting by the millwrights. This pushed the electrical installation back 48 hours and so the electrical contractor had to work 12 hours on the actual holiday at triple time.

Although the old-line start-up after shutdown went all right, there were innumerable problems when operations tried to start-up the new equipment. A new gearbox had no lubrication and promptly gave up - no spare in stock. Some process change parts were lost somewhere and others arrived with shipping damage requiring last minute fabrication at a local machine shop.

A production employee was injured attempting to unjam the new carton conveyor the first day of start-up requiring a shutdown for safety modifications. Although well instructed by the process engineer, operations demanded lots of changes to the former so they could operate and clean the system effectively. Contractors and technicians had to work around the clock the following weekend to execute the changes. Even so, the first 5 attempts at production runs of the product ended with 5% acceptable product and 95% scrap.

Ultimately, the new line extension product was available off the revised production line 5 weeks later than desired. Marketing had to pull advertising and reslot it for 6 weeks later in one region. Start-up costs were 230% higher than budgeted. End of month 10 accounting report $495K allocated, $5K unallocated - and the project still needed $133K in additional funds for still to come contractor bill submission for field changes, unanticipated overtime, and extra material costs.

The PM had to go back to the business for an additional $128K - much to the business' displeasure. Two months later another bill from the packaging change parts vendor arrived for $5K in additional engineering and remanufacture costs because of the carton size change - it had been lost in their accounting department. The $5K was charged off to another project.

Later that year the project manager was rated "needs improvement" in his annual performance review and his step increase was withheld. Six months later he decided to leave the company. The following year the marketing director vetoed considering the site for another new product since that plant had a "less than stellar record" delivering on commitments.

Project Bravo: The project scope was also well defined: add some new raw material handling items, and process items; provide change parts to forming equipment; provide change parts to the packaging system. Get it running with minimum start-up costs and make product for distribution to coincide with the marketing launch plan. Essentially it was the same as Project Alpha.

After getting all the product configuration data from research and packaging data from marketing, the PM set up a detailed budget, withholding $50K in a PM's Contingency account. She met with her team to explain expectations, giving each team member a copy of the budget and a detailed write up of what she expected to have happen by when. She insisted that one line maintenance packaging technician and one senior operator were assigned to the team, even though these employees could only serve part time.

Operations wanted some old equipment in the process area removed and a new storage area built ($30K). But she told them that the project could be tight and that if there were any funds remaining she would consider it once the project scope had been met. The designers assigned to the project felt like they needed some additional software to speed the piping design. She discussed this with the Engineering Manager and they agreed to have a team look into this - but to fund it, if recommended, from local small minor project funds.

Detailed budgets were set up, short weekly meetings were held, and a running "what has been spent - what is yet to be spent (re-estimate)" MS Excel spreadsheet was developed by the PM. Each assigned employee got a copy the day after the weekly meeting. Monthly the group reviewed the accounting report also.

Initially the team got together as much as their busy schedule would allow to review designs while they were still just computer images to work out interference's, look for potential operation, cleaning or maintenance problems, and agree on progress. During one of these meetings a significant process flow design mistake was spotted and quickly changed. Maintenance provided their standard list of spares carried in the stockroom and all agreed to specify equipment that took advantage of theses spares.

Process equipment was specified, and ordered. Process piping design started. Electrical design awaited process and packaging design completion. Packaging change parts were specified and ordered with an 8-

month lead-time. Things appeared under control. The monthly engineering report indicated on time - on budget. The second month's accounting sheets showed $200K allocated, $300K unallocated.

Marketing then asked for the carton configuration to be changed from 24 count to 12 count. Sales had advised that 12 count would move faster and appear like product was turning over on the stores' shelves faster. This doubled the number of cartons to be handled per unit time and required a redesign of change parts by the vendor. About the same time, the maintenance technician assigned to the project found that the cartoner had been modified.

The PM and packaging engineer decided to call in a service engineer from the vendor to review the current machine condition and help engineer the parts to fit. This required approval from the Division Manager since the business was very careful about allowing outside vendors into their plants. Other machines were covered and the visit had to occur on a Saturday. The packaging engineer and PM then met with the maintenance technician, operations, and the vendor's representative and redesigned the carton conveyor. There would be additional cost so the PM released $20K from her contingency to packaging for the up charge. End of month 6 accounting report: $350K allocated, $150K unallocated.

Work continued. Process piping design was a bit late - the designers blamed it on antique software. When issued for bid, process piping came in 5% higher than estimated. The PM and Commercial met with the successful bidder and worked this down to 3% by coordinating with the electrical engineer on what work could be done when. They also met with other concerned engineers and operations people to coordinate all work scheduled for the shutdown - to insure that contractors would not be asked to perform more tasks than they possibly could in the time allotted.

This showed the need for a very detailed installation plan and this was developed over a long 4-hour meeting by the whole team. This meeting lasted until 7PM and operations was not happy about picking up the overtime for the senior operator and maintenance technician.

Then a vendor for one of the key items of process equipment advised that there would be a delay in shipment due to a potential strike at one of his vendor's factories. With a significant effort, Commercial and the PM, working with the vendor, managed to find a back up source for this material, and ordered it with a 25% restocking charge if there was not a strike. The PM covered this cost with $5K from her contingency.

Raw material samples were late and when they arrived the viscosity of one was twice the original - necessitating a pump capacity change. Electrical advised that the new pump needed all new wiring to meet code. The PM met with the electrical engineer and they worked out a way to save a little in one area to cover the costs in another. But she still had to release $5K from her contingency to cover the up charges.

The operations supervisor assigned to the project was transferred to another site and a new person was assigned. The PM brought this new person up to date and asked him to meet with the senior operator and maintenance technician to work out an operator training plan for the new equipment.

Electrical bids were 5% higher than estimated. The electrical contractor refused to reduce his lump sum price but agreed to do the work on a "time and material" basis. He anticipated that he could bring it in close to the original estimate, if everything went well, but did not want to take the financial risk of a lump sum contract.

Once he was shown the detailed installation plan, and met again with the process-piping contractor to coordinate activities, he relented and accepted the lump-sum contract. The team still had to make up the 5%. But the packaging engineer had been able to save a little with the packaging change parts vendor. The strike did happen but the contingency material worked well and the vendor was able to keep to his original parts production schedule, so the restocking charge was not needed.

The team gave the stockroom a detailed list of what material and equipment was to be received when, and who to contact upon its arrival. Each day of the last week prior to installation material on site was reviewed and purchasing expedited missing items. End of month 9 accounting report: $400K allocated, $100K unallocated.

A short 4-day holiday shutdown was scheduled for installation. Later this would reduce to 3 and a half days due to high production demand. Installation work commenced. The PM was on site off and on most of the shutdown, along with her team. There were still last minute changes. When one became obvious the team and the contractors would meet quickly and work out a solution. Even so, they had to authorize some additional overtime for the electrical contractor.

The old line start-up after shutdown went Ok. The packaging engineer and maintenance technician carefully ran cartons for the new system by hand down all the conveyors to check for pinch points, while the process engineer and senior operator inspected every fitting and pump to insure they would clean easily. The maintenance tech spotted a new gearbox that had no lubrication and filled it. Some process change parts were lost in shipment and others arrived with shipping damage requiring last minute fabrication at a local machine shop.

Operations demanded some changes to the former so they could operate and clean the system effectively. After a meeting to discuss these, two potential safety items and one cleaning item were considered critical and two technicians had to work Sunday, the following weekend, to execute the changes. The demand for other changes went away, as the operators became more familiar with the system. The first five attempts at production runs of the product ended with 45% acceptable product and 55% scrap. The last run of the five was 90% acceptable.

The new line extension product was available off the revised production line as desired. Start-up costs were 93% of budget. End of month 10 accounting report: $470K allocated, $30K unallocated. Contractor bill submission for field changes, unanticipated overtime, and extra material costs came to $15K. So $15K remained. After a discussion with the plant manager, the $15K was coupled with $15K from the operations repair budget and the old equipment operations had requested was removed and a new storeroom installed.

Later that year the project manager was rated "meets expectations" in her annual performance review and slated for a developing special project to improve her visibility to the business. The site was identified for a new product in the 5-year business plan. At the fall trade show one of the attendees noticed that the cartoner manufacturer had incorporated the reliability changes executed by maintenance on the site's cartoner into their new model cartoner offered for sale.

So what is the difference between Project ALFA and Project Bravo?

There are many. And we will discuss each. But the key element is that Project BRAVO was managed well. Project ALFA was managed somewhat but mostly just allowed to happen. Project BRAVO was executed by a team, working together to help each other. Project ALFA was executed by a loosely coordinated group of individual contributors.

But most of the differences are subtle. Let's examine them.

Organization: Project Bravo was very well organized. Operations and Maintenance hourly employees were really included in the team. These people *know* what is there now and the tiny details of what will need to happen to get a new system running. They live with the mistakes of previous projects every day. And they want this one to be more successful. The PM used their expertise and experience.

Every person on the team received a paper from the PM which stated the objective in some detail, established the overall and individual budgets, set out milestones and dates, enumerated the requirements for communication, and talked about start-up. Each team member felt like he/she had an *important* job to do individually and still knew that they would be involved in the decision making process along the way. They were *committed*.

They knew that there would be many meetings and that they would each report on their progress at each meeting. They felt *responsible* for the project as a whole, not just their individual piece of the pie. The importance of setting up a team working environment from moment one in any project can not be over emphasized. Without it, things fall through the cracks. With it you at least have a chance of catching most of them before they fall.

It is true that the PM for Alfa held a meeting - one - to kick off the project. The difference is in the depth of knowledge transferred, the depth of expectations presented and the depth of team responsibility insisted upon. Project Alfa became another pain in the environment that the team members had to give time to. Project Bravo was an opportunity to play on a well coached team with a purpose.

Financial accountability was emphasized in Project Bravo. Although the accounting department has to keep score on commitments made, and report monthly, this information is like the monthly statement from your bank on your checking account. If you didn't keep track of the checks and debit card withdrawals you made between bank statements you could very easily end up with a negative balance. Your bank statement has items that are 5 weeks old on it - and is, at the best, two weeks ago worth of history. In addition, the bank statement shows what was spent - not what still needs to be spent. It makes a great tool for balancing the books and helping the company tax people get the deductions and depreciation the business needs - but is not much good at keeping score for a project manager.

A PM needs to have a way to keep score daily or weekly and to use total commitments plus anticipated future commitments in *aggregate* to know where they are and where they are going. Setting up a simple MS Excel spread sheet, insisting on a copy of each purchase requisition issued, spending or having a clerk spend a moment or two a day updating data, disseminating the sheets often to the team and insisting that each team member re-estimate commitments yet to be made, insured that the Bravo team got very few financial surprises.

Another aspect of financial accountability was the way the Project Bravo PM resisted the irrational exuberance that pervades the approval of a moderate to large project. "Wow, look at all the money we have!" "We must have some extra money in all that - let's do one or more of those things we have wanted to do for so long and just couldn't find funding for." "It's close enough to the project scope gray area. Let's do it."

But the Project Bravo PM knew that many other project financial troubles have been caused by this irrational exuberance. It is a classic mistake. She knew that the business world is a dynamic place and that over the course of the project there would be unanticipated changes and problems to deal with. And these take money, sometimes lots of money. She refused to even consider unessential expenditures from the project until the end - and had to resist heavy pressure from others because of that.

She set up a $50K (10%) PM contingency that ONLY she could disburse. The remaining $450K was parceled out to the team members with responsibility for their individual portions of the project based on discussions with them and review of the detailed project estimates. She knew that the PM contingency would be spent - very carefully. A bit of underestimation here, a couple of higher than anticipated bids there, some unanticipated overtime - all things that she was expected to be able to deal with, and now prepared to do so.

You shouldn't infer that if you do all the things discussed under financial accountability above, you will not get into financial trouble on a complex project in a rapidly changing environment. But it does mean that you will see it coming long before you end up in deep: time enough to bring it to the business's attention. Not fun. But essential and the earlier you tell the business that everything is not coming up roses, the better for the business and you. A business does not like bad news. But a business hates late, beyond salvage, you should have told me long ago, bad news.

Getting into financial trouble on a project is a mistake. Getting into financial trouble and not knowing (or refusing to believe) it is incompetence. Big difference. And yes, some project managers will stand on a sinking financial ship with water lapping on their feet and expect a miracle to save them. "Maybe the installation bid will come in real low, maybe I will find a money tree, maybe just maybe ..." They all drown.

Other Items: Making a unilateral decision to update some software prior to it being used for an important project is like sticking your head into the lion's mouth and praying he decides not to chomp down. Everyone knows that a software update does three things - fixes bugs in the previous version; adds more bells and whistles that the software company's marketing arm will use to peddle more software; and adds a whole new set of bugs. But upgrade you must to stay current and maintain support. The trick is to manage the upgrade. Do it incrementally across computers being used for that purpose. Do it once the initial issue has been out there awhile so others find the bugs and the software company has time to fix them. Buy it when service pack one is included. And do it when you can easily cope with the new bugs.

A classic example is Windows XP. It wasn't until Service Package Two, almost a year after the initial issue that the operating system wouldn't go off on its own and present the computer operator with the blue screen of death. A whole lot of data and work was lost to Windows XP. We all know better than to trust a software upgrade to do what it is advertized to do without a glitch. Project Alfa ignored this and bought an upgrade.

Electrical design generally has to wait until process or packaging design is almost complete. The electrical engineer has to know what equipment will be where, how much energy it needs, and more importantly, the sequence of operation. A delay in process design almost always delays electrical design. With today's computer systems and programmable logic controllers, lots of the electrical software design can wait a bit.

But not forever. Code takes time to write and time to debug. The total time is about 50/50. Project Alfa got behind this eight ball when process design was late. And it is always true that the longer you take to get the electrical engineer the data he/she needs to do their thing, the more bugs there will be that you do not find until start-up.

Whenever any installation that uses funds from a project is being installed - especially on a weekend - someone from the project team with the authority to make decisions, or at least privy to the PM's cell phone number, should be present. It was the project's money and therefore the project's (and the PM's) responsibility. Project Alfa let the floor installation go in without anyone there. The operations manager who authorized the extra should have not done so - but the project Alfa PM could not un-spill the milk on Monday. And to a contractor, any supervisor is the company's representative and they will attempt to please.

In addition, it is almost always true that a verbal order by a company supervisor to a contractor is a binding legal contract, regardless of whether the supervisor was authorized to make such an order or not. Someone from the Project Alfa team, well briefed on what was to be done and what was not to be done should have been on site during the floor work. Then the extra work would not have happened.

The request from marketing for a carton count decrease from 24 count to 12 count was a significant speed bump to the project. This was a big budget risk. The entire project team should have been involved. Revised designs and estimates were required. Just calling a vendor and telling them to "do it", authorizes the vendor to spend whatever they feel is necessary to "do it", and all this is an extra to be billed on a time and material basis plus overhead and profit. So the vendor's contract was now wide open. For a packaging engineer to just call a vendor and say "do it", violates an engineer's responsibility to have purchasing involved in change orders.

Then Project Alfa did not estimate the cost of the revised carton conveyors. So the project Alfa team didn't even know how much of a speed bump the Marketing request was. Project Bravo did. And understood the complexity this added. That is why they elected to bring in the vendor's engineer - not a cheap decision ($3K), but a good one even with the additional approvals and another Saturday away from more enjoyable weekend activities.

Of course the cartoner manufacturer's representative took the information on the maintenance reliability changes back to his own engineering department - and used it. Customer modifications are always of interest to equipment manufacturers.

Sometimes bids come in higher than estimated. Lots of things control this. Material or labor price increases or decreases, how busy the contractors are, the precision and completeness of the bid documents, how much of a material discount the contractor can get from his suppliers, how much confidence the contractor has in the engineer, maybe the phase of the moon. If a contractor has to source skilled labor from outside his normal crew, it will cost more.

Cajoling a contractor to reduce his bid is not good management. Every contractor knows that anything but the straightest forward all new, no tie-ins to existing equipment project will uncover problems during installation and they will require extras that the engineer has to authorize. So cajoling just moves costs from the firm lump sum portion to the extras portion - it does not reduce it.

The right way is to understand why the contractor's bids are higher than anticipated and then, working with the contractor, reduce the contractor's scope and risk. Good, reputable contractors appreciate this. Take a look at what the Project Bravo team did. They coordinated all contractors' work for the shutdown so that the contractor felt more comfortable (less risk). Then he was willing to reduce the contingency he had placed in the bid.

Being advised that a vendor's supplier might go on strike is a normal occurrence, in some countries far more normal than in others. Ignoring it will not make it go away and the vendor may interpret your actions as an indication that time is not of the essence, i.e.: you don't care when he delivers.

But working with him to find another source of supply, and the willingness to put some dollars on the line for contingency shows him you are very serious about delivery. Project Bravo handled this correctly. When the strike happened, the secondary source delivered even though the primary could not. No restocking charge.

Raw material samples from a vendor's normal production that are different from the preliminary laboratory samples are normal. Expect it. Every project should be able to roll easily with these kinds of punches.

So also is having project team members change during a project. The business is dynamic; people move around to take advantage of promotional or broadening opportunities. Personal lives change. The Project Bravo PM anticipated it, and dealt with it to insure a minimum disruption to the flow of team work.

There is an attribute of good project managers that we called "forehandness" previously. This is the ability to spot potential problems before they become apparent to others. It is not clairvoyance. It is keen perception. Most problems do not just jump up, full grown and ready to devour the team. They have a gestation period. And sometimes they are sired by a lack of motivation or responsibility on the part of a team member - generally exhibited by a bored, "it's not my job" attitude.

Others are vendor related. Early indications of delayed delivery dates indicate some internal vendor problems. Still others are functional related. A new product launch without a test market to establish the basis of sales volume may well mean a last moment change in product configuration as marketing gets more focus group data. Look closely for them. They should trigger an intuitional tic in a good project manager's mind. Don't ignore the tic - investigate it.

Never accept a "Time and Material" bid for anything but the smallest items. This is an open ended contract where the contractor has every right to charge you his total cost plus overhead and profit. And it is in his best interest to inflate his cost. Good, reputable, trustworthy contractors will play reasonably fair. Others will not. There is no motivation for a contractor to figure out how to do things cheaper on T&M. The Project Bravo team handled it correctly. Reduce risk and get a firm lump sum contract. Even so, field changes and extras, which will occur, will be necessary.

In every project the team has to track the delivery of material and locate it for quick access. Losing something isn't very good management. And following up on deliveries that are late is purchasing job. They do it well if they know.

Shutdowns cost the business lots of money. And shorting a customer's order is not good business. So most shutdowns shrink. Expect it.

A shutdown installation takes an enormous amount of coordination of every detail. You can not have two of your contractor's occupying the same space at the same time. Electrical always has to wait for process or millwrights to finish setting equipment prior to connecting it. A glitch on day one reflects instantly into a delay on day three. Many PM's work the process and millwrights day shift and electrical night shift.

It costs a bit more for shift differential for the electrical boys, but it precludes them climbing over one another. The team has to be on the floor with the contractors most of the time. There will be problems and they need solved immediately. Having a bunch of millwrights standing around waiting for a decision costs money and a contractor will want paid for it or be less willing to just absorb the cost of the next glitch.

There will be field changes required and the PM and engineers need to discuss them with the contractor, negotiate a not to exceed price, and get a change order issued the next time Purchasing is on site.

Every piece of new equipment needs checked for lubrication, free turning, and safety prior to turning on the power. The sequence of operation needs checked. Expect the requirement for an outside machine shop to, on an emergency basis, fabricate a part or two that were unanticipated or damaged in shipment. Stuff happens.

Always prepare a start-up plan indicating what you intend to do, when you intend to do it, what you hope to learn, what raw materials and labor you will use, and what will cause you to shut down. Review this with the team and operations before installation and each time before a test run. Until the new system makes acceptable product on a sustained basis, the system is "yours" - it belongs to the project team when the new system is to be run. Don't run for hours making scrap. Shut down and fix it. Start-up management is the project manager's responsibility until operations accepts the new system as "theirs".

After the first test run all systems seem to have more problems than they will ultimately have once the operators know how to run it - something they acquire through experience. Making every change operations wants initially is just an opportunity to ultimately change it back the way it was once the learning has occurred. Fix EVERY safety item immediately. Fix all quality items immediately. Fix nice to have stuff later. But making the determination of what to fix is the job of the project team working closely with others. And if you have established a rapport with operations and other interested and effected parties from project inception, this is a whole lot easier.

There have been many technically superior projects that became failures because the operators did not care to go the last extra mile to make them successful. You have to establish a rapport with the people who will do the product assembly or run the production line. They can make a poor design run well and a great design fail miserably.

Since there are always extras during installation, and always some essential changes during start-up, a good project manager insures that he/she has the remaining resources to cover these items.

There is an old question: *When is a project completed?*

a. When the project scope has been met

b. When the money is gone

c. When the engineer gets assigned to other projects

d. Item b. and c. above.

Too often the real answer is d. This is not right and leaves a very bad taste in the mouths of the operations people on the receiving end of projects. They live with the less than complete results forever. The engineer goes on to other things.

The right answer is "a", but this does not mean everything is perfect. It is functional and delivers what it was suppose to deliver. *No business can afford 100% projects where every tiny detail is 100% perfect.* It costs just as much to get from zero to 95% as it does from 95% to 100%. So you have to manage the effort to get to 96% or 97%. Then, over time, the operations people can tweak it to attempt to approach 100%.

So we are back to the beginning. A well organized project, with a lot of people involvement, team work, resource accountability and responsibility, and careful management will be successful. Those missing these items will not.

Synopsis: *A well run project has the following characteristics:*

1. The project team is well organized and every team member **understands** what needs to be done, by who, when, with what resources. The team feels "ownership" for the project and its results. There is *one* project manager with overall accountability and responsibility for the project.

2. *All departments* with a stake in the results of the project are involved; especially those knowledgeable about the details of what hardware, software, and operational procedures already exist. Operator and maintenance *training plans* are developed and executed before start-up.

3. The team *meets frequently* to review designs, progress, and evaluate projected costs. All team members know what is happening and appear involved and motivated.

4. The team *communicates effectively* with other aspects of the business involved in the overall business venture that the project is supporting.

5. Timely and *accurate financial tracking* is in place - what has been spent and what still needs to be spent.

6. There are some *contingency funds* held in escrow by the project manager to deal with potential problems.

7. The *entire project team addresses changes and problems* when they occur. The team

tends to spot potential problems early and contingency plans are developed for potential problems.

8. There is a *tracking system* for incoming material and a way of expediting delivery if necessary.

9. There is an effective *installation plan* so that all work will be coordinated and workflow will not be interrupted.

10. There is an effective *start-up plan* and a clear understanding of what must be achieved for the team to turn the project over to the operations group.

There are other items also depending on the project size and type: major expansion, greenfield site, developing world, technical development, etc. - but the above ten points are the *key* elements to the successful management of any project.

Small Projects: Successful small projects have the same characteristics, although there is nowhere near as many people involved, and the project manager may be the only management member on the team. He/she is then his/her own engineering staff. Small projects are normally managed concurrently with other project responsibilities. A engineer may have one medium sized project they are project manager for, a couple of medium sized projects they are assisting other project managers with, maybe a large project they are on the preliminary engineering team for, and a few to many small projects they are executing all at the same time and all at different stages of completion.

Keeping track of multiple project assignments requires discipline, attention to detail, and persistence.

Large Projects: Successful very large projects also have the same characteristics. And they are always managed as a well coordinated and integrated group of smaller projects, with an overall project team and sub project teams. The subprojects are again broken down into subproject elements. So there are subproject managers and subproject element managers reporting to subproject managers. But there is still only *one* overall project manager.

Typically there are multiple engineering disciplines involved. Sometimes there will be external engineering firms doing a portion of the project. (The effective use of external engineering firms will be discussed in detail later.) A very formal Critical Path Schedule or PERT chart is prepared and maintained. Most project resources are full time, and some may even be from other sites or a division or corporate engineering staff. Communication and coordination is critical and a communication plan is developed and executed. Continuous construction site supervision - generally by one or more designated "owners" representatives (project team members) are in place once construction starts.

Large projects require a significant effort to manage well and only a very experienced, competent project manager should be assigned.

It would be very difficult to define a standard organization for a very large project since they are all different and what requires emphasis on one may not require the same level of emphasis on another.

Lets look an example of the development of a large project. The following is written as a combination narrative, and as a short story. It tries to convey the human side, along with the technical. Although it is not a reflection of any real event, it is a narrative about what can really occur in the preparation for a large project.

Engineering Management, an Irreverent Primer

Section III, Chapter VI

PROJECT DEVELOPMENT

A Short Story: Project CHARLIE: A major consumer product business was happily experiencing sales growth for a core product. The five year business plan called for adding some line extensions to this successful product to broaden its market footprint, while concurrently providing advertizing and promotion efforts to maintain or accelerate the core product's sales. Although the business had about 20% excess manufacturing capacity at current sales levels, capacity would become an issue very soon. A significant flexible capacity addition would be essential and since additions of this assumed magnitude do not happen overnight, Project Charlie was initiated to look at doubling (+28,000 metric tonnes) the capacity of an existing plant on owned land next to an existing structure. We will call the plant site Smalltown.

With the amount of capital that was to be involved (a preliminary guess was $30 million) a pre-project for Preliminary Engineering was approved initially and this pre-project was anticipated to require 6 months and $500K. If the large project was approved the pre-project would be capitalized. If not, it would have to be expensed.

Since a large building was going to be required, attached and tied into the existing plant structure and sharing utilities, civil and architectural engineering, and major utility engineering would be necessary. This was in addition to the process and packaging assets necessary.

This organization was a division of a large international corporation. Divisions were powerful, and had their own President and top management staff. Yet divisions were held responsible and accountable for delivering profit to the corporation, ensuring sustainable growth in sales and profits and expanding market share.

The goal was for the pre-project manager to do the pre-project, manage the preparation of the project write up, and the presentation to top management. He/she would also be the prime candidate to ultimately manage the final overall project. However, this was not guaranteed. After a lot of discussions between the divisional technical director and the division staff, a project manager was identified and appointed.

His name was Alex. Alex had demonstrated, with larger and larger projects his ability to deliver. Alex was 34, married, one two year old daughter, an engineering graduate of Podunk University with a degree in Mechanical Engineering. After three years with another consumer products company, he had joined the firm five years ago as a project engineer and had two years ago, been promoted to senior project engineer. He had served in a one year cross functional assignment in operations, and then relocated to his current site on a lateral move. He was indentified as an employee with the potential to take on more responsibility (promotable in the future) - but only after he had demonstrated his ability to deliver on a large complex project. However, he had never been in responsible charge of anything of this magnitude. The divisional staff was betting on an unknown.

Since Alex was on the division engineering staff and his home was near another site (Middletown), he knew he would have to accept a temporary assignment and would have to live out of a suitcase - at least for awhile. This was an "opportunity" for Alex. And like all opportunities, a risk. He spent a long difficult weekend discussing it with his wife after the divisional technical director had presented the idea to him.

It would mean sacrifice, and probably a temporary or even a permanent move again. It would mean that he would be traveling even more than he now was. It would mean some very long hours - especially if the big project was approved and he ended up project manager for that. It would mean less time with family. But it would also mean that he would at least be in line for promotion and a significant salary increase. If he did a good job.

Crunch time. Where did he want to go with his career and where did his family want him to go? Support from his spouse would be essential and a decision to accept this assignment would mean that she would be alone a lot - managing the household and taking care of the daughter.

Her career aspirations would, at best, be delayed. How could she continue to take a few courses towards her Masters Degree now? She had taken a leave of absence from her middle school teaching job when her daughter was born. It ran out soon. Was she willing to risk losing the ability to easily go back to the profession she loved?

A difficult choice for each and for them together. Once he committed, and if he reneged later, it would spell the end of his upwardly mobile career with this company. This was a crossroads. They both knew that.

By Sunday night they agreed that he should take the opportunity and that they would live with the consequences. It was a tough decision. He called the technical director Monday morning and accepted.

Once the pre-project manager was announced, the opportunity to participate in a large project with high visibility elicited many telephone calls and emails recommending specific individuals to the team. The band was playing and almost everybody wanted to be in the parade. Everybody had a favorite son or daughter to push

Staffing the team was not easy. Naturally, Alex wanted the best. After 5 years with the business, he knew a lot of people and their capabilities - or at least had heard of their capabilities from others. But sorting out who was the best, who was willing to participate given the personal sacrifices involved, and who was available was not easy.

Some good people were already halfway through other important projects and the division was reluctant to pull them out now. One well known capable engineer had announced that she was going to have a child - which meant that if the project was ultimately approved, she would be unavailable, on maternity leave, for the first critical few months of the project. Not good. One had just accepted a lateral developmental assignment to another business unit in the corporation and changing this was not in the cards. He had to accept one engineering member from the site even though they were a relatively new, untried, and inexperienced group. Frustrating.

Around and around Alex went. Finally a list of the team primary and secondary selectees started to gel in his mind. It took Alex, with a lot of help and interference running by the divisional technical director and in one instance even the divisional manufacturing head, to finally sort it down to a first and second choice short list. He was then authorized to "make the offers".

Alex went to division headquarters (Bigtown), and the different sites, presented the opportunity to the primary selectees, and waited. Roughly like putting a coin in a slot machine and waiting for the dials to stop spinning to see if he won or lost. He got three out of five the first draft. Then he went for the second choice and got two for two. Not bad. His team was selected: a raw materials handling engineer, a process engineer, a packaging engineer, an electrical, and a utility engineer. A good group that, with some coaching, had the knowledge and experience in total to deliver - even if a couple were a little on the inexperienced side.

But one of the second choice people could be a problem. Although highly recommended by the engineer's boss for her outstanding and innovative technical skills, she had indicated that the engineer could be less than a team player. She was very dedicated but had little time for those who did not share this dedication. Her boss indicated that her interpersonal skills left a bit to be desired. Was this an understatement? He would have to watch out for her running roughshod over peers and subordinates. An exhausting exercise but he finally had a team! Had it only been two weeks? Seemed like ten.

Getting the team together for the first meeting at Smalltown was another challenge. No one in an effective organization is just sitting around waiting for the fire alarm to ring. Each team member was involved in other things and those things had to be sorted and the responsibility transferred to others. Other site plant managers had issues. It all required persistent sorting. But ultimately it worked out: two weeks later.

Alex filled the two weeks at Bigtown, with discussions with other functions relating to the range and type of line extension products, and a detailed review of the site and existing utility capacities, He sat down with the divisional industrial engineering manager to get preliminary production capacity needs on paper. A staff industrial engineer was assigned to the project bringing the total internal project headcount to seven. This was a large positive. IE input was essential. After reviewing things with the divisional technical director, manufacturing head, and the rest of the divisional top management group, Alex was ready to commence work on preliminary engineering.

Alex then went to Smalltown for what turned out to be four days of almost non stop meetings. He met with his team. His team met with the site management. They toured the proposed new plant location. They reviewed the past history of the project idea, where it came from, why it was important, what they had to do, by when, with what resources. Getting them to mold into a team was not going to be easy. They were all competent and capable - at least to some degree - but they were also strong individuals that would require work to coalesce.

A dinner out together, an evening silly sports event where the team played the site's operations managers (the project team lost and Alex was sure *not* the star player for his team) was worth while. They were getting to know each other. And they were getting to know Alex and what was expected and what was not. A team doesn't spring instantly into life - it grows.

Since Smalltown operated on a "just in time" ship to regional distribution centers, no additional finished goods warehousing was necessary. Raw materials were a different matter and both bulk storage facilities and a palletized warehouse would be required. Industrial engineering would size these elements.

The preproject team again met at Smalltown and started conceptual design work. Some very rough layouts and utility need estimates were made. Job shop designers were hired to work with each of the key engineers. At $50/hr for 6 months the cost estimate was $132K total. This seemed excessive and the team agreed to cut back on designers as soon as feasible to try to limit the cost. But for now, they needed hard copy to discuss.

It looked like a 7500 square meter single story wide bay structure would be required. Since the division and corporate did not have any registered professional civil, architectural, or structural engineers on its staff, an outside consultant firm was necessary.

Alex and two of his assisting engineers reviewed the credentials of five firms, and requested proposals from three. The request for proposal called for five separate conceptual building designs, two of which would be selected for design development, and one of which would be selected for a detailed construction cost estimate. Alex and his team were spending four out of every five working days at Smalltown, flying in on Sunday night, and maybe getting back home late Thursday night to work independently at their home site on Fridays. Tough on the social life.

The team ultimately settled on one: A.C. Engineers, Inc. - a firm with a lot of consumer products' experience located 200 km from the site. They had recently executed preliminary engineering for another firm (not a direct competitor) which led to the construction of a new facility in the same area. They seemed to understand the requirements well, and had a staff of experienced and solid professionals. Since the team did not have expertise in heavy electrical power, the consultant engineer agreed to look at that also. A.C. Engineers agreed to assign a project manager from their staff and a team of architectural, structural, and electrical engineers. It was not cheap. The initial consultant engineering contract was $200K.

A.C. Engineers required a site and topographical survey with elevations to benchmark on a 10 meter square grid. A set of soil core samples at key locations from the site was also necessary. This cost $80,000, and Alex was concerned. He had only been project manager 6 weeks and had already committed $412K or 83% of his budget!

After a discussion with his team, he agreed, on a trial basis, that the on site schedule would be modified to allow remote people to arrive by noon on Monday and return home on Thursday night - working independently like they had been at their home site on Fridays. This allowed for more "home" time and kept them off the crowded, generally delayed airplanes on Sunday nights. With the corporate intranet, video conferencing, and the telephone they could keep in touch.

From a personal viewpoint, Alex continued to be at Smalltown early Monday (red eye Sunday night) through Thursday night - although this was really starting to cause lumps in his family life, and his sleep account was somewhat overdrawn. But he knew he had to establish a rapport with the Smalltown management group if he had any chance of being successful with Project Charlie. Those were the people that would have to live with the results of this project for years and they needed input and a relative comfortable level that could only come from knowing and trusting Alex.

Core samples showed low soil bearing ratios at the preferred building location, so the planning had to include much wider foundations than anticipated if the preferred location was to be used. In addition, the topographic survey indicated quite a bit of site overburden movement would be required to improve storm water drainage and provide for proper perimeter access. Existing underground fire protection lines and sanitary sewer lines would have to be relocated. The guestimate of $30MM for the actual project was in jeopardy.

At this stage, and until a final project had been approved, the division manufacturing head decided that a formal announcement to the local community and site employees was not to be made. Site employees were told that the site was being considered for expansion but that no firm plans yet existed.

Even so, the Smalltown plant manager and Alex met with local governmental officials to determine permit requirements and the ability of the local infrastructure to cope with additional truck traffic. Site wastewater flow would also increase and the existing municipal treatment facility was insufficient. This then added a requirement for waste water pretreatment at the plant prior to discharge. A.C. Engineers was issued a $40K change order to have their waste water treatment group participate in this portion of the preproject. Now Alex was up to $452K - or 90% of his project budget. Sleep started to come less easily, even the night after a red eye.

The utility engineer and the electrical engineer met with local Smalltown public utility representatives to insure that sufficient electrical power and fuel would be available if needed. This appeared to be OK.

Concurrently, the project team engineers reviewed raw material handling, processing, and packaging options. They visited other company sites that had executed major expansions over the past few years. They attended an international trade show to check on new innovations or ideas that may be applicable. They discussed the merits of developing new technology in-house vs.: applying current technology with some evolutionary refinements. They prepared many different plant layouts, and discussed the pros and cons with the entire team and Smalltown operations management.

There was a lot of controversy on this item. Some indicated a desire to go for the revolutionary - a step change in technology to improve productivity. Some argued for the "tried and true" to minimize risk. The division technical director was called in for a meeting to discuss the most valuable proposals and the risks associated with each. At the end of a long day it was decided to take a middle path and try some new high potential items while leaving enough space and flexibility to revert to the "tried and true" if necessary. They also met with the consulting engineer regularly to coordinate activities.

Once the A.C. Engineers proposed its five alternatives for the building, the team evaluated each very carefully. Two were considered good, one was marginal, and there was very little difference between the other two and the rest. Two were selected for additional design development work. Team engineers then concentrated layout work on these two building options. With some considerable effort, production line alternatives were boiled down and evaluated.

Ultimately, the project team considered two significantly different conceptual designs as viable alternatives, each with a couple of sub options. Preliminary cost estimates were prepared for all of them. The consultant engineering firm took care of the site work, structure, and major utilities, including waste water. The project engineers took care of utility distribution, raw material receipt and handling, process, packaging, and transfer of finish goods to the shipping point.

Then, industrial engineering developed staffing plans for each while the project engineers evaluated potential energy use, and incremental costs for increasing capacity further in the future. Technical risk was also evaluated based on the blend of revolutionary and evolutionary development required for each option.

The following evaluation chart was prepared.

Project Charlie

Alternatives

Alternatives #	Capacity Available MT	Capital Cost $(000)	Time to Comp MFAP	Increm Staff #/shift	Energy Used MJ/T	Increm Capacity 10KT $(000)	Technical Risk L,M,H,VH
1	30000	31650	18	18	2100	5100	H
1A	32000	32125	20	14	1900	4500	VH
1B	32000	29875	18	20	1950	6200	M
2	28000	27450	18	18	2250	8500	M
2A	27500	28100	20	16	2150	4890	H
2B	26000	25625	16	24	2300	9340	L

MT = metric tonnes at full capacity

MFAP= months from approved project

#/shift= incremental operations & support staff per shift for the site

MJ/T= Megajoules of energy required per tonne at full capacity

L,M,H,VH=Low, Medium, High, Very High

This type of chart is typical and helps management understand the trade offs. Different options deliver different results. Some have a higher initial capacity. Some save capital. Some take less time to implement. Some are more automated and require less operators and support staff, some are more energy efficient, some are easier to expand in the future, and some have less technical risk associated with new equipment designs or new processing techniques. Preparing this chart was the project manager's responsibility. It gives the top management of the business *options*.

They can have their new capacity in larger of smaller bites. Upside capacity needs longer term may be more or less expensive. They can have it sooner or later. They can have it a bit cheaper but sacrifice productivity. They can have it very productive but at a cost in time and capital. *Everything is a trade off.* And it is almost always true that "bigger" and "better" or "faster" costs more and carries higher risk.

They were now ready for a presentation to top management, complete with computer generated plan and elevation block drawings and some computer generated perspectives of the proposed structures and production lines. The canned presentation was targeted to last 90 minutes. Alex insisted that they do a rehearsal first at Smalltown and got the plant manager, and division technical and industrial engineering managers to "act" as the critical audience. Some aspects were revised, some deleted, others added. The team made up a list of expected questions and answers. They were ready.

Alex and his key engineers made a formal presentation the next Friday morning at Bigtown to top management. The Smalltown plant manager, the divisional technical director, engineering and industrial engineering managers attended.

Then there were a lot of discussions of the alternatives and costs. And some very different points of view. The brand leader for the core product wanted all the capacity possible right now. The financial group wanted the most capacity for the least money. The operations people wanted the highest productivity. The presenters were questioned often on how the numbers for each option were developed and if there were other feasible options that could be considered, and what other options were looked at and why they were rejected.

What top management was looking for was the best option and a comfort that the project team had reviewed all feasible options and carefully boiled them down to the best. The large project would gobble up 18% of the available division's annual capital cash flow and preclude or delay some other items they would like to do to grow the business in other areas. And until the sales anticipated from this additional capital investment materialized, it would put a visible dent in the division's return on invested capital and asset turn over ratio.

It was a big bet on the future. There was also some concern that the 5 year plan capacity needs may have been based on some wishful thinking on the part of the sales group. Finally, recent competitive action in the marketplace indicated that the launch of the line extension products proposed may be more difficult than anticipated.

The project team had had an insightful, if frustrating, opportunity to see top management decision making in action.

At the end of the meeting Alternatives 2 and 2A were selected as the best of the group presented. But one was not approved. The project team was directed to go back and refine these a bit further. Top management really liked the lower cost, slightly lower capacity, and lower incremental cost for future expansion of 2A. They did not like the higher risk and the longer time. What they asked for was 26000 tonnes of capacity for $26 million in 16 months, 16 additional employees per shift, 10,000 incremental in the future for $5 million, and moderate technical risk. That is, they *wanted to have their cake and eat it too!*

The meeting lasted through lunch, and ended mid afternoon. The project team was tired, hungry, and faced a minimum one hour car trip and a 2 hour airplane ride to get home, best case. One of the team members would not get home until almost midnight.

The project team left the meeting discouraged. They thought they had done a super job and now there was even more work to do. There was some grumbling about "would the team ever come up with something acceptable?" "Did top management really want to make a decision?" "Was sending it back to committee just a way to delay making a decision they were not yet prepared to make?" Alex felt a deep frustration and burn out in himself and his team. They had built up to the big meeting over the last few weeks and now they felt like they had played their heart out and still lost the game or at least were down 21 points at the half. It was obvious that cheer leading on his part right now would be resented. His team needed a break.

As they departed Alex thanked them for their effort and praised the excellent presentations and how well they handled being on the "firing line" when top management was hurling difficult questions. He told the team that he had confidence (did he really?) that they would come up with an acceptable alternative. He told the team to work on their own to come up with ideas at their home site until Tuesday, and that he would schedule a team meeting for late Wednesday morning (so their travel could be early Wednesday instead of late Tuesday night) at Smalltown.

On the way to the airport Alex and the packaging engineer discussed options. Alex thought out loud, "If a lot more refinement and alternative evaluation is going to be required I will need more money in the preliminary engineering project." They both agreed to touch base on Monday, and Alex called travel on his cell phone and arranged for a flight for Monday night. He planned to spend Tuesday going back over the staffing plan with the local plant manager. He also needed to take a hard look at the preliminary engineering project budget and try to find some places to squeeze out a few dollars.

It was a good weekend at home, even though there was a bit of tension due to his travel schedule. Mulling over options to try to meet top management's request occupied Alex's mind a lot when he had a spare moment.

On Monday morning, at the Middletown site, he got a telephone call from the division technical director. The technical director said he thought the presentation went well, and complemented Alex and his team for the work. He said that after the meeting he was talking with the division operations head who also liked the presentation. The division president dropped by and also had praise for the team's effort. The president pointed out that the day before, he and the division finance head had received a telephone call from corporate. It seems that another division, overseas, was having some major distribution problems due to competitive action. They would not be delivering the profit corporate expected this year and maybe next. Corporate had a team of distribution experts headed over there now to try to help with damage control. But, all other divisions were being asked to cut back on spending. *A business is just like a balloon - poke it on one side and it bulges on the other.*

This was probably going to be a 20% capital budget reduction bulge! Significant. The president indicated that he had told his management team this Friday morning. So they went into Alex's presentation knowing there was going to be a cash crunch yet they had not had time to work out the details on how they were going to reduce spending, including capital expenditures, and yet still not totally disrupt their growth plans. So the top management group was in no mood to agree to anything, especially a $30 million ($15 million a year for two years) capital project.

The president also indicated that the top management group knew that something had to be done to increase capacity for the core product. So Alex had been asked to cut costs and deliver more.

Alex told the technical director thanks, and promised to touch base again by Friday. Meanwhile he would get to work on seeing how close they could come to what top management wanted.

He also asked when the technical director thought the division top management would sort out the priorities and provide some guidance on just what projects were going to stay in the capital budget and what were going to go. The technical director said that the division operations head had called a plant manager's meeting at Bigtown for Tuesday to get operations perspective. The rest of the top management group was collecting information from their staffs. The top management team was meeting with the president on Thursday. So an answer would come soon.

As Alex was hanging up, he thought, "I can't meet with the Smalltown plant manager on Tuesday, he will be in Bigtown for the plant manager's meeting. I'll go to A.C. Engineering on Tuesday instead and get them cooking on peeling some cost out of the building structure and site utilities." Alex knew that there was only *one* effective way to reduce an estimate - *take things out of it or replace high cost materials for lower cost materials.*

You just can not say make it cheaper and not be prepared to make the tradeoff's necessary to make it cheaper. *More with less doesn't work at the detailed estimate level.* It might be fine for top management to say do it, but something has to go to make the estimate come down. Otherwise, you will have an overrun. His job, working with his team, A.C. Engineering, and the Smalltown plant manager was to figure out what had to go.

He also called all his team engineers, explained what the technical director had told him, leaving out a little of the why, but insuring that they understood that there was a capital crunch coming and that everybody had to pull together to squeeze the budget. He also told them that the divisional president had been pleased with their presentation. He asked each of them to come to Smalltown for the Thursday meeting with their list of items from their area of responsibility that could be cut or reduced. Alex started working on his own list of potential candidates for the ax.

By Tuesday night, after a half day with A.C. Engineering, Alex had a list of about 30 items, none of which he liked. Some were drastic - like making the building structure insulated metal panels instead of insulated tilt up concrete panels. This would save quite a few dollars but metal panel buildings were a lot less durable and expansion in the future would be more costly since metal panels did not remove and relocate well.

Some were not so drastic, like reducing the floor area that would be covered by acid proof brick to only the essential wet processing areas, and going with epoxy over concrete in the other processing areas. Not the greatest for long term looks and maintainability. Not the greatest for future wet processing expansion, but a nice savings to the budget.

Wednesday morning Alex met with the Smalltown plant manager, and his local industrial engineer. They first discussed the previous day's plant manager's meeting. It looked like the plant managers had come up with a few capital projects for site replacement work, like a new roof for a warehouse, or a boiler replacement that could be safely delayed.

Then they got down to reviewing the proposed staffing and discussing options. This was very difficult. Jobs are a group of tasks. You may be able to reduce a person's workload, (take away tasks) but *you can not reduce a part of a person*. You have to get to one whole person. You have to shift tasks around, consolidate pieces of work and try to get the consolidated pieces to equal one whole job. And people can not be in two places at the same time. You can look at increased automation, but then that increases capital cost. You can look at support staffing, like maintenance and clean up. But at what risk? And the more changes you do to save money in capital may well require more support staffing not less. Capital is a one time cost. Staffing is a forever cost. The only way to fairly evaluate this financially is to compare the life cycle of the line using a discount rate to get a net present value, and then compare that with the incremental capital cost.

Since the team had done a good job with staffing the first go around, there was very little slack to take up. This was going to be very, very difficult! Finally, the plant manager told the industrial engineer to get with the individual shift supervisors for the existing lines and see if there was some way the new line could share some support staffing. Alex was not hopeful that there was going to be any significant staffing reduction possible.

On Thursday the entire team met, went through everyone's list of potential cuts, and ultimately ranked them in order from the least disruptive to the original design and proposed operation to the most disruptive. As would be expected, if they changed something in raw material handling, this required a change in processing. If they changed something in processing, it required a change in packaging, etc. *Everything was connected to everything else.*

By the end of the day the group was floundering. They had more questions than answers. More details had to be worked out. Some of the designers they had originally had been let go, and they needed them back. The preproject budget was approaching the red line rapidly. They agreed to keep working Friday and Monday, and to meet again Tuesday at Smalltown.

Alex put a late call into the technical director. He was in a meeting. On Friday morning the technical director called back. Alex told him about the budget crunch. He agreed to fund the additional designers and any up charge from A.C. Engineering for their additional work from his budget for now, and Alex was to prepare a small overrun request to bring the preliminary engineering project up to $550K. As soon as the overrun was approved (it only had to go as far as the division president) any charges would be journal vouchered back to the project. He also indicated that the top management group had come up with a few marketing related items that could be delayed, and that the capital budget crunch would be more like 12%. That was very good news.

Alex wrote up the overrun request, got the plant manager's signature, and faxed it to the technical director. Then he headed for an airplane home.

Monday was a home site work day. Alex called the Smalltown plant manager and was told that with a significant effort the plant would agree to a 2 person reduction per shift, one from operations and one from support, if and only if the packaging automation proposed for alternative 2A remained in the budget.

Weather conditions were not conductive to air travel into the Smalltown area, so the team met Tuesday at noon via video conference. Again the work was to sort the cut priorities, this time with a lot more understanding of the tradeoffs and interrelations between different items. The alternative 2A packaging automation was taken off the table. It would stay in the budget. It was a long video conference - and another group who had planned to use the video conference facility at Middletown and Bigtown had to be asked to reschedule - which they reluctantly did.

Alex assigned each team engineer to work up a detailed estimate based on the team's agreements, closed the meeting, and went to the phone to A.C. Engineering. The team had taken out some acid proof brick, opted for a larger electrical power transformer to reduce cost for expansion, went to epoxy coated block instead of ceramic tile block for some interior partitions, and modified the heating, ventilation and air conditioning (HVAC) scheme to roof mounted package units in non critical areas - reducing chilled water piping costs. There were many changes to raw material handling, process, packaging, utilities, and controls to achieve what they were looking for. The tradeoffs were not what Alex would have liked given more available funding, but the structure would be functional, sparse, but functional, and the production system would be effective and robust. The revised estimates were due Thursday.

On Thursday the team met in Smalltown. Everything was reviewed and agreed. This is what the team came up with where Alternative 2C was the new alternative they had developed.

Project Charlie

Alternatives   -
Second Generation

Alternatives #	Capacity Available	Capital Cost	Time to Comp	Increm Staff	Energy Used	Increm 10KT	Technical Risk
	MT	$(000)	MFAP	#/shift	MJ/T	$(000)	L,M,H,VH
2	28000	27450	18	18	2250	8500	M
2A	27500	28100	20	16	2150	4890	H
2C	26000	25825	18	18	2330	6110	M

Top division management had asked for 26000 tonnes of capacity for $26 million in 16 months, 16 associates per shift, 10000 incremental in the future for $5 million. They had hit the tonnage and cost, missed the timing a little, and missed the staffing and incremental addition cost. But they had done a whole lot better than Alex had originally anticipated.

Alex called the division technical director and told him what they had. Two hours later he called back indicating that the division operations head was in agreement, and that Alex (alone) was to bring all the data and present it at the next management team meeting the following Tuesday in Bigtown. Also, the division president had signed the overrun request for the preliminary engineering project.

Alex had an excellent weekend. The following Tuesday morning he presented Alternative 2C to the top management team at Bigtown. It was essentially a non event. There were only a couple of clarification questions and he was told, "Proceed to write up the official request for capital project approval" (which had to be signed at division and go to corporate for approval.) So with the exception of adjusting the format of some drawings for inclusion in the project write up, the preliminary engineering project was completed (unless corporate threw them a curve ball).

Alex called the Smalltown plant manager, and then his team with the good news. He gave his team some instructions on the preparation of project write-up information, and told them to release their designers as soon as possible. He then called A.C. Engineering, and gave them final instructions. The next day he picked up a copy of a couple of approved large capital projects from the file, and spent the rest of the day with marketing and research and development at Bigtown getting them started on their portion of the project write-up. Alex was home early on Friday night. Monday was a holiday so the family could have 3 days together. Wonderful, and needed.

The Project Write-up: Since a project write-up is essentially a **business proposition** and **not** a technical dissertation, Alex needed some help. Consultation with the Smalltown accounting manager and the division finance director was necessary followed by a lot of work with division industrial engineering.

Alex's project was very complex from a business viewpoint. The new capacity was going to be used for both an existing product and line extensions to that existing product. As stated previously, business propositions involving capital expenditures are evaluated over 10 years. This means that Alex had to gather Profit and Loss projections for the existing product **and** the proposed line extensions for 10 years and then allocate line production time each year to each product.

Then he had to come up with a consolidated Profit and Loss for the new capacity. Each Profit and Loss statement had to include the estimated gross sales value of product produced on the production system. Each Profit and Loss statement had to include estimated costs for raw materials, labor, advertizing and promotion, distribution, fixed costs, repair and maintenance, cleaning, overhead, energy, start-up, taxes, and all the other things that have to be paid for before profit is obtained.

And each year would be different as the line extension products were added and as the incremental 10,000 tonnes was added. He then had to calculate an Internal Rate of Return over the 10 years for this consolidated Profit and Loss Statement. This was a major crystal ball exercise.

Once that was done, he had to develop a sensitivity analysis, and show the IRR (Internal Rate of Return) for each. Like what happens if sales are 10% short, if raw material costs increase, if the project overruns, if one of the key technical risk items fails to deliver, etc.

This was a significant task and required a lot of help from accounting. For a good project write up, Alex had to be not only an engineer; he had to be an accountant, an industrial engineer, and an author. But most of all he had to be a business person, not just a technical person. He had to think like a business person, use business terms, and present the project as a business opportunity that required a capital expenditure.

So, over the next three weeks, Alex, with a whole lot of help from others, wrote the project. The first few pages were almost strictly business proposal. There were multiple Appendices that included the proforma Profit and Loss statements, a narrative about the assets to be added in Smalltown, and drawings showing the site location, building footprint, building perspective, production system layout, process flow diagram, and a discussion of technical risk items. What would be required to execute a future project to expand capacity by 10,000 tons was included. Another appendix discussed a proposed project organization and implementation plan. Still another discussed start-up plans and cost.

Although the business proposal was only 6 pages, the overall project write-up, with Appendices, was 36 pages. Alex's first draft went to the Smalltown plant manger and the division technical director for review. Concurrently, he had the draft reviewed by marketing, sales, purchasing, and finance. Some changes were required.

After the changes were incorporated and agreed, the finished project original was signed by Alex, the Smalltown plant manager, the division technical director and operations director, and the chief industrial engineer. The marketing manager responsible for the product line signed also. It then went to the division top management team. As expected there were questions and Alex had to spend a few days in Bigtown answering them. Finally the division management team, followed by the division president signed, and the project was sent to corporate.

With the travel schedule of the corporate staff, it took awhile there. The division president had to answer some business plan questions. There was some concern by corporate finance that the sales projections were a little aggressive. There was a request from the corporate manufacturing staff to provide a side letter on what the project manager intended to do to address fire safety in the new building in more detail.

In most businesses, top corporate management uses *levers* to keep control of the business. These levers are:

1. Approval of divisional expense budgets
2. Approval of long term plans.

3. Approval of short term plans.

4. Major capital project approval

5. Approval of the appointment of senior managers.

These levers are very effective. Budget approval keeps expenses in line. Plan approval allows corporate to have some course control on business objectives.

Major capital approval gives them the opportunity to review what the division really wants to do to expand the business and coordinate cash flow corporate wide. The major projects submitted can be compared with the long term division plan and the short term division plan to insure that the division is following up well on their approved plans. Once a major capital project is approved by corporate, it is real money and corporate will hold the division accountable for spending it wisely on what was listed in the scope of the project, and obtaining the business growth and profit anticipated by the project.

There will be an audit, most of the time at the one year and three year point. Not that corporate will send someone to do a detailed audit unless the project was obviously mismanaged. In that case there will be a bit of a witch hunt culminating in some unhappiness for many people. Otherwise, corporate will expect the division to audit itself and report.

The other lever, maintaining an advise and consent role in the appointment of senior managers, gives corporate the ability to help select candidates and approve appointments of the people who will actually be running the business at the director level and above. Corporate can present their own candidates from other divisions insuring broad experience prevails. Consequently corporate top management not only knows the people selected but can groom the high flyers for even more responsibility in divisional and even corporate executive positions.

Success for Alex. Finally, *the project was signed by top corporate management*. This was the go order. After a lot of preliminary work, $550K in expenditures, and almost 7 months, *the money was really there*. Now the challenge was to deliver on what had been promised - functional capacity, on time, on budget. It had taken a lot of work to get to this point - and this was just the beginning. Eighteen months from now the new capacity needed to be in place and fully operational. Today all that existed was some grass and dirt. A big challenge.

The short story of the birth of project Charlie is not unique. It is typical. Getting approval for a significant capital expenditure is almost as bad as getting legislation through congress. It is an iterative process and the manager stewarding it has to be technically competent, politically savvy, knowledgeable about the business, and have a good reputation across broad areas of the company.

For an engineering manager, selecting the right person to be project manager for large projects is critically important.

Section III, Chapter VII

ENGINEERING STANDARD SPECIFICATIONS

Every project engineering group should have a set of Engineering Standard Specifications. They become the boiler plate of your bid packages and save time since your engineers are not engaged in reinventing the wheel for each project. Most businesses make these standard specifications at the divisional level. When bid packages are prepared the applicable standard specifications go out with the project specific specifications and drawings to insure vendors are bidding with more complete knowledge of what is required.

Standard Specifications consist of:

General Conditions – Applicable to All Work. This is the real boiler plate and should indicate that a contractor is bidding on "providing all material unless otherwise specified; supervision, labor, scaffolding, tools, sundries, and all things necessary to procure, receive, warehouse, deliver, install, construct, erect and test all items required to totally complete the work." The contractor should have a competent foreman or lead person on site; store and secure material in a designated area, set up his/her site office and fabrication shop in a designated area, and follow standard OSHA regulations. You will have to flesh out the General Conditions with help from purchasing.

Special Conditions – Every business has some special considerations for coping with lots of contractors on site during construction. Things like the designated construction gate, areas where contractors are restricted from entering; cafeteria hours for contractors so that they do not overload a cafeteria during normal worker meal periods, special invoicing procedures, who has authority to authorize change orders and how they will be handled, etc.

Then you get into the nuts and bolts. You should have a standard specification for those things that your business does a lot. A food business will have one for stainless steel process piping. A heavy industry may have one for wood block floors. Most will have one for utility piping or electrical conduit installations. Some will have one for asphalt paving; some for concrete installation and finishing. Some will have standard specifications for surface preparation and painting including the site's standard color scheme. Each business is unique.

Your engineering standard specifications should not be static. They should develop over time and as technology advances. They should list the manufacturers that your group has chosen as the ones who provide the material your business has selected as acceptable. A special note: the addition of the words "or equal" opens a closed specification and places the responsibility for insuring that the material in question is equal on the contractor. This doesn't work real well and can get you into a finger pointing session with the contractor. Using "or approved equal" forces the contractor to get the engineer's approval prior to substituting material from another manufacturer.

Standard specifications should be sufficiently detailed to insure that you get what you want. If you require stainless tubing to be welded in an inert gas environment to insure a smooth weld interior, say so and indicate that some welds will be cut open to insure that this requirement is met. If you require a certain dry thin film thickness for applied paint, state that you will use a gauge to check this. Then do so. Saying that you will check something then not doing it sends a "we don't really care" message to your contractors.

Don't fill specifications with references to ASTM (American Society of Testing Materials) numbers unless you are sure that the bidders fully understand the reference. Most contractors estimating staffs are familiar with the normal ones, but for exotic items you must include a copy of the ASTM sheet right in the specification. And stay away from exotic material wherever you can – it costs far more than what is available off the shelf to contractors and they have no purchasing leverage with exotic suppliers.

Having a good set of engineering standard specifications speeds the bid package preparation process and insures that contractors understand the standards they will be held to.

Engineering Management, an Irreverent Primer

Section III, Chapter VIII

ESTIMATING

Estimating capital project costs is a black art. It takes experience and luck. Get it wrong on the low side and you have a guaranteed overrun. Get it wrong on the high side and you may well price yourself out of the market.

Almost every business has rules of thumb for capacity additions. Usually it is an average cost per unit of production capacity like X thousand dollars per hundred units produced per year. This is fine for ball park estimation. And top management knows this rule of thumb and will question you extensively if your final estimate exceeds it.

Projects that are not straight forward capacity additions don't follow this rule of thumb but top management will expect them to unless your estimate is below it. Anticipate their questions and have good understandable answers.

Although we all know that an estimate is just that – an estimate - a best guess based on incomplete information and a little Kentucky windage. But once an estimate gets incorporated into a capital project write up and the project gets approved it is no longer an estimate – it is a budget. You, or someone on your staff, will be expected to deliver against budget.

The key to estimating is knowing what significant items have to be purchased and what work has to be done. And you can't know until you have done, at the very least, a preliminary design. The better and the more precise your preliminary design the better your estimate. Do as good a job on developing a preliminary layout and major equipment list as you can or have time for. Identify the technical risks since they are inherently items that generate costs far above standard.

There is only one way to produce an estimate that is iron clad. You must essentially do all the detailed engineering, produce detailed drawings, specifications and equipment lists, get actual vendor quotations for equipment, get firm lump sum bids for construction and installation, and then add ten percent to the total. Any other way entails risk of overrun. Almost no one can do this. It takes too many resources and way too much time. So you end up doing the best you can in the time allotted and managing the risk.

All good estimates have the following characteristics:

1.   They lay out costs in a standard and predictable format and are broken down into sections.

2.   Each section has a line item for "Miscellaneous Unidentified." This is where the small items that are not captured in the main part of that section of the estimate are included – typically 10% of the section's sub total. It is NOT a contingency – it is a line item for all the small parts and small work that you know will be required but you have not taken the time to specifically identify every one.

3. Each section is broken down into purchased equipment and installation costs as separate line items. Usually purchased equipment costs are easier to obtain since purchasing can call a vendor and get a budget price on most items. Budget prices are usually higher than actual bid prices. Installation costs are a lot harder to estimate. A good local contractor may be willing to give you a budget price but in most cases you have to come up with one yourself.

4. Sections are then added up and a contingency line item is listed. Contingencies are for unexpected costs and are an integral part of the estimate – not some figure that can be taken out and held in reserve by some other manager. You will spend the contingency. You just don't know on what yet. If your business insists on some upper manager holding a "contingency", then call the contingency line item "Overall Project Miscellaneous Unidentified", and then add another line item for contingency. Then don't plan on spending the super contingency.

5. Historical data is a wonderful source of costs – if inflated to current value. If someone in your group or division has done a similar project in the past, pick their brains. Most good engineering organizations keep a data base of every project estimate and the project final accounting for every large project done across the division.

6.   Estimates are based on actual items to be purchased and installed. If the estimate is to be cut, actual items to be purchased and installed have to be taken out; i.e.: the scope of work has to be reduced.

One format for estimates: This is an example of an estimate for a project to build a structure and install a consumer products production line at an existing plant site:

Estimate: Project Delta, Fruitville site, Dated _____

1.0 Engineering

1.1 Internal Engineering	$xxx
1.2 External Engineering	$xxx
1.3 Consultants: Site Survey	$xxx
1.4 Consultants: Soil Samples/Analysis	$xxx
1.4 Miscellaneous Unidentified	<u>$xxx</u>
Total 1.0	$xxx

2.0 Mobilization … (getting the required manpower on site and setting up any work and office space they need, rigging temporary power, water, etc.)

3.0 Site Preparation …

4.0 General Construction: Building …

5.0 General Construction: Utilities Service …

6.0 Process Systems

6.1 Process Equipment	$xxx
6.2 Process Piping Material	$xxx
6.3 Process installation	$xxx

6.4 Process utility connections $xxx

6.5 Miscellaneous Unidentified   <u>$xxx</u>

6.6 Total Process Systems $xxx

7.0 Packaging Systems

   7.1. Packaging Equipment

     7.1.1 Depalletizing $xxx

     7.1.2 Container Cleaning $xxx

     7.1.3 Filling & Capping $xxx

     7.1.4 Pasteurizing $xxx

     7.1.5 Labeling $xxx

     7.1.6 Cartoning $xxx

     7.1.7 Casing $xxx

     7.1.8 Palletizing $xxx

     7.1.9 Miscellaneous Unidentified   <u>$xxx</u>

     Total Packaging Equipment $xxx

   7.2 Packaging equipment Installation

     7.2.1 Line front end $xxx

     7.2.2 Line back end $xxx

     7.2.3 Carton/case conveyor systems $xxx

     7.2.4 Miscellaneous Unidentified.   <u>$xxx</u>

     Sub total Packaging Equip. Install $xxx

   7.3 ……   <u>$xxx</u>

   Total Packaging Systems $xxx

8.0 Electrical

   8.1 Power distribution …

   8.2 Control Systems …

   8.3 ……

9.0 ……

10.0	
11.0 Project Misc. Unid.	$xxx
Sub Total Overall Project	$xxx
12.0 Contingency at 10%	$xxx
Project Total Cost	$xxx

If you develop a system somewhat like this (the line items are not important nor are the sequences although 1.0, 2.0 and 3.0 are normally so stated) and use the system for every estimate your group produces, upper management will get used to it. They will assume, hopefully correctly, that your estimates are reasonable, sufficiently detailed without being so detailed that it takes forever to produce one, and well thought out.

There are a lot of formal estimating systems out there. You can adapt one to your needs. The above is one I adapted from a combination of the American Institute of Architects format and the RS Means Cost Data books. I adapted it in a way to allow the different engineering disciplines preparing the estimate to have their own estimate for their portion of the project – what will be their own budget for the work to be performed if the project is approved.

Here is another example for a large ($7MM) commercial project: a 130 unit senior apartment complex. The format is a bit odd. Numbers are $(000):

1.0 General Conditions, OH & Profit	$736
2.0 Permits	$ 25
3.0 Performance Bonds	$ 70

4.0 Clearing, Mass Excavation	$257
5.0 Erosion Control	$ 15
6.0 Storm Drainage System	$ 85
7.0 Landscaping	$135
8.0 Asphalt Paving	$ 90
9.0 Sidewalks	$ 60
10.0 Concrete curbs & Paving	$ 55
11.0 Sanitary Sewer Systems	$ 90
12.0 Water Distribution Systems	$ 85
13.0 Concrete Foundations	$166
14.0 Gypcrete Floors	$ 45
15.0 Concrete Slabs	$153
16.0 Brick Masonry	$320
17.0 Stucco	$164
18.0 Structural Steel/Metal Stairs	$ 43
19.0 Framing Labor	$280
20.0 Lumber, Panels & Sheathing	$309
21.0 Floor Joists	$ 90
22.0 Roof Trusses	$100
23.0 Finish Carpentry	$247
24.0 Cabinets & Tops	$115
25.0 Thermal & Weather Protection	$400
26.0 Doors & Windows	$126
27.0 Finishes	$664
28.0 Specialties	$ 60
29.0 Appliances	$ 75
30.0 Elevators	$ 55
31.0 Plumbing	$802

32.0 HVAC System	$455
33.0 Electrical System	$596
Total	$6,968

This estimate is in bank format. In fact it isn't an estimate at all, it is a budget. It is broken down like a bank would want to see it for a construction loan. Each line item can be separately reviewed in the field during construction as to the completeness of each line item for the authorization of draws (payments) to a general contractor. The disbursal of money from the item 1.0 General Conditions, OH & Profit is based upon the combined value as a percent of the total project of the other individual line items completed at the date of the draw inspection. Some banks will require a 10% holdback on each line item until completion just to focus the contractor's mind. So a contractor will inflate his draw request. It's all just a game.

You will note that there are no "Miscellaneous Unidentified" items. Banks don't like them but you can bet they are buried in the individual line items. There are also no "Contingency" items. Banks hate them. So they're also buried.

Normally a bank will contract with a separate outside construction inspection company to visit the site each time a contractor requests a draw and see if the contractor's percentage of completion for each line item makes sense. Most inspectors will agree with the contractor unless something is glaringly overstated. And material delivered to the site, but not yet in place counts towards completion percentage. Many contractors have material delivered just prior to inspections. Then they have thirty days or so to pay the material vendor and may be able to get a draw prior to doing so. It reduces their internal financing costs.

Cost data books and data bases: They are very valuable for estimating building construction costs and have information on local areas usually by city and include both detailed, work crew, and square foot costs. They are great at showing building cost inflation over literally decades. They are not good at estimating costs for production lines, processing systems, and specialized items. And they tend to cost out buildings using building trade's rates (union) and work rules. If you are constructing a building in an area where building trades do not prevail, costs will be different.

Good engineering managers have developed a consistent system for estimating that fits the needs of their business.

ESTIMATING PHILOSPHY

If every estimate you or your department produces results in projects that always come in on budget and never result in an overrun when the project is complete, your group is over estimating. That is they are spending more in aggregate for the buildings, systems and production facilities they are asked to estimate and build. We all know, or at least should know that a project's cost will expand to fill the budget – that every dollar authorized will be spent if the project manager is not so tight he or she squeaks. Some one will always come up with another bell or whistle that they desperately need if they know there is a dollar left in the project budget.

Perfection would be that out of every one hundred estimates you produce 50 of them overrun slightly and 50 of them come in slightly under budget when finished and the unspent dollars are turned back in as unneeded. Then you have some assurance that you are estimating well and not overspending for the facilities you install.

But business doesn't work this way. If you come in on budget no one says much. It is what you are expected to do. If you need to go back and ask for a little more money top management wants your first born son as a token of your contrition. I tried for years to get top management to understand this creeping overspending phenomenon. I was only moderately successful. Requests for overrun approval always cause grief.

BACKING INTO AN ESTIMATE

There is an even more insidious type of overspending. I call it backing into an estimate. If your company's capital project hurdle rate is say 25%, then it is possible that a capital project estimate can be inflated to produce a 25% to 30% return on investment even if a good tight estimate would be lower and produce an ROI in the 40% or higher range. This may be especially true if the project is for additional capacity for existing products at a newer site that has had recent expansions.

Very seldom can systems be selected and installed that exactly match the production capacity targeted. Equipment capacity is a step function. Example: If you need 25,000 pounds per hour of saturated steam you may have to install a 30,000 pound per hour boiler – they come in 20,000 30,000, and 40,000 sizes. So when that project is done you have 20% excess steam capacity.

Over time at a continuously expanding site you get to the point where additional incremental production capacity may be had very cheaply – just for some minor debottlenecking. Yet the backed in estimate says the site still needs almost every dollar it would need if it was starting from scratch to expand capacity. It is a gross overcharge.

Resist the tendency to back into an estimate. It is doing your company a major disservice, gobbling up capital for idle capacity, and possibly shortchanging other projects that may help the company grow and diversify. It can lead to facilities that are far fancier than they need to be to produce a quality product. And it inflates your asset base and lowers your ROIC.

Section III Chapter IX

EXTERNAL ENGINEERING FIRMS

The use of external or consulting engineering firms is usually a requirement of large projects. Your organization is not normally staffed with the numbers or the specialized disciplines essential to do preliminary engineering work or to execute an actual large project. Usually the specialized disciplines are Architectural Engineers, Civil Engineers, Sanitary Engineers or specialized Utility Engineers. You just don't have enough work to keep these people busy all the time to keep them on your permanent staff.

There are lots of consulting engineering firms in the United States and many in overseas locations. They range from huge firms that could build a ten square mile residential island in Dubai to small ones that do small offices in your home town. But they basically do the same thing. They do preliminary engineering and final engineering and construction management for the capital projects of others.

If they are successful and have been around awhile they are good at what they do. If they weren't, they wouldn't be in business. Most of the middle sized firms specialize, at least somewhat, in the industry they serve. Some serve the energy industry (oil, gas, coal), some the process industry (chemicals, ethanol, pharmaceuticals), some food, some transportation (trucks, automobiles), some infrastructure (rail, roads & bridges), and some aerospace. Some do mining and tunneling. Some do schools and universities. Each is quite capable in their nitch and each is staffed with people experienced in their areas. They are competent. But they are not cheap. If they bill by the hour for their people, expect to pay a competitive salary times two to two point five.

You can get them to do anything from engineering studies, full blown preliminary engineering, cost estimates, detailed engineering and bid package preparation, equipment procurement, construction contractor selection, construction management, and total responsibility (conception through start up) work.

A sideline to this is something called "design build", sometimes called "turn key". Design build is usually done for stand alone items like warehouses, tank farms, generic utility packages, etc. You bid it on a performance specification: how many pallets you want to store, what the acceptable temperature range is for storage, how many gallons/liters of liquid you wish to store, etc. Then you select the bidder who appears to provide the best value for money. The selected bidder does the design, bids construction, purchases the material, and manages and completes construction. In the end you get a key and a bill.

There is a caution that needs stated for design build and turn key. The performance specification has to be very detailed and specific. The design build firm will cut corners and has the incentive to do so. With a lump sum bid every nickel he/she can save on material or construction is a nickel in his/her pocket. You will get lower quality roofing materials and bare minimum insulation. The interior air distribution will be barely sufficient. Concrete will be bare minimum structurally and may get poured when environmental conditions are bordering on too cold or too warm. Structures will be to code – but just to code. Do not be surprised if the facility looks great initially but requires a lot of maintenance. Five years later you may be in for extensive and expensive repairs. I have seen a 150,000 square foot design build warehouse require a total new roof membrane five years after it opened. That's almost six football fields of built up roofing – not cheap.

The comments on design build are not to degrade the engineering firms in the design build business. It is the nature of the beast. They provide a specialized service that some companies want. You will get a sparsely functional facility for a very low, relatively speaking, initial cost. But the life cycle cost of the facility will be higher than a conventional project where the project's owner (engineering staff) is in the loop every step of the way from designing through final completion. You pay up front or you pay incrementally. It's your company's choice.

When you select an outside engineering firm you first have to select the three to five firms from which you want to request proposals. Most engineering managers start by looking at firms with demonstrated capabilities that match the project. Then they visit the firm's offices, look at their past work, talk to satisfied customers, and get a review of their financial status (Dunn and Bradstreet financial review).

The next step is to select the firms that will be asked to submit proposals. Here is where the performance specification comes in. You have to write a detailed performance specification on the work you want them to do. Don't scrimp on the effort you or your team puts into this: The better the performance specification the better the proposal. The performance specification has to include all those things you would do if you were doing the job yourself – including timing. How many formal progress reports do you want? Where – at your offices or theirs? What information will you provide to them? Remember, every bit of data you give them they will assume came down from the mountain on stone tablets. If you give them bad information you will get a bad result.

While the engineering firms are preparing their proposals, visit some of the facilities where they have been involved and talk to their customers. In most cases these visits will be set up by the engineering firm, so don't expect to be sent to dissatisfied customers. You will need to dig for them on your own. Research Engineering News Record – the tabloid of the engineering industry.

When you get the proposals back do not be surprised if they are very different from each other. Some will be polished like the annual report of Exxon/Mobile. Some will be a little rough. What you are looking for is content, not flash.

Most proposals are presented in a meeting. Usually they come to you. Insure that you have a good multifunction group of your own people present including your assigned purchasing professional. They should have a list of questions that each of your people attending understands, that is:

Does the presenting engineering firm understand the performance specification totally? Are they taking exception to one or more items in the performance specification? Why? Note: this is not a negative – in fact it may be a significant positive.

Have they done some preliminary layouts and concepts?

Have they identified who their project manager is going to be and not only presented his/her credentials but hopefully have he/she present for the proposal presentation?

How about their other employees who will be involved?

Have they developed a time line? A set of milestones? A plan for progress review meetings?

Have they developed a cash flow projection? How much will they want paid and when? Do the draws match the milestones?

Have they presented examples of previous work they have done for others? Are these examples relevant? Recent?

Have they developed a preliminary schedule of site visits for their employees that will insure that they understand the details of tie ins to existing structures and systems involved in the project?

What do they need you to do? Do they need a site survey, topographic survey, soil compaction tests, underground utility locations, existing utility excess capacity studies, etc.?

Have they presented a "guesstimate" of overall project cost? Is this even close to your own "guesstimate" and if not, why not?

What rates will they charge for principles, project managers, individual engineers, designers, clerical personnel, etc. if there are extras? Note: there will always be extras.

Are these rates reasonable? Usually the rates are in the range of two point five up to four times salary. So if an engineer would get say $100,000 per year and the benefit multiplier is 1.75, then the base is roughly $90/hour and the billing rate is $225 to $360 per hour. Senior engineers are more and it goes up from there. Project managers ... principles; you can see billing rates above a thousand for principles.

After you have received all the proposals it is evaluation time. Get your team together. Review the questions. Remember that cost is always a consideration but not the only one and surely not the most important one. You are hiring a consulting engineering firm for their experience, competency and capability. They will be working closely with you, your team, and others in your organization. Do you feel comfortable with them and especially the employees they will assign to the project?

Do a formal evaluation listing pros and cons for each. Then do the final selection, notify the winner, send a very nice thank you letter to those not selected and get purchasing to prepare a formal contract. Once a contract is signed by both parties you are ready to begin.

The stages are normally:

1. Engineering Study. This is where the firm determines your needs, develops concepts, prepares preliminary flow diagrams and layouts, develops better "guesstimates" and comes up with alternatives. They will expect you to select an alternative that most meets your needs. Usually one or two alternatives need tweaked a bit.

2. Preliminary Engineering. Here the selected alternative is fleshed out more. A more detailed conceptual layout is made. Building construction types are firmed up. More detailed internal alternatives are developed. Ball park estimates are prepared. Note: a ball park estimate is not ready for a project write up – it is better than a "guestimate" but not good enough to commit.

3. Design Development: With relatively firm capacities, flows, and conceptual layouts in hand the consulting firm can now prepare more detailed layouts, equipment lists, structural loading and final building footprints. From this a project estimate can be prepared. Usually this is sufficient for you to write the formal capital project and hopefully get approval by your top management to proceed. Don't be surprised if there isn't a lot of back and forth tweaking here. And a tweak with a consulting engineering company is an extra.

4. Final Engineering: Here the consulting firm prepares the detailed drawings and specifications necessary for permits and the preparation of bid packages and equipment purchasing lists. At the end of this phase you should be able to quote and

purchase equipment and get contractor bids for the construction and installation. Normally a consulting engineering firm will help with quotation and bid evaluation. This phase takes the longest and is the most expensive.

5. Construction Management: Once contracts are issued the consulting firm will manage the construction and installation of the project. They will have one or more construction managers on site and will work with the contractors and sometimes suppliers to execute the project. They will hold periodic construction meetings to work out details and preclude conflicts. They will prepare punch lists of work that needs done or redone. They will act as construction quality control managers.

6. Start-up: Once the building is complete, equipment installation is complete, and the facility is ready, they will perform the preliminary start-up - checking out equipment, control systems, and making sure everything is ready mechanically and electrically for raw materials. They will then assist you in the actual start-up of the facility and the first production runs. They will assist in determining essential changes and longer term change considerations. Some will train your people in the operation of the facility.

7. Post Project Support: Once a project is complete a consulting firm will normally be available, for a fee, to consult with you on changes, expansions, alterations and major repairs if you so desire.

Do you always hire a consulting firm to do all the above? Seldom to never. You hire them to do individual phases. Many companies might hire a consulting firm to do an Engineering Study and Preliminary Engineering, relying on their own in house staff, usually augmented with individual contract engineers and designers ("job shoppers") to do the remaining phases. It should be noted that there will still be some residual involvement of the consulting firm at their normal billing rates to help in the transition between their people and yours.

Other companies will have the consulting engineering firm do the first three phases or even the first four. Many times a consulting engineering firm's scope of work will be limited to buildings and utilities only and the actual process and packaging engineering design, and installation will be done in house.

Every project is different and every company's internal capabilities are different. Consulting Engineering firms understand this and most are willing to do everything from a full seven phase project to just advising or doing a portion of a project.

You must manage a contract with a consulting engineering firm with at least as much supervision you would provide to a building contractor adding a room to your own home – maybe more. There will be changes in the scope of work. Upper management will want to tweak things. And everything outside the original contract performance specification will be an expensive extra. In most cases you will require a firm to prepare a few alternatives and then you will select one for design development and estimation.

If you have a good rapport with the firm's project manager you can preclude problems before they get out of hand. Don't be afraid to just stop everything where it is and have a high level meeting with the firm to sort out problems. Remember that once a consulting engineering firm gets started on a project the meter runs faster than the diesel pump filling an eighteen wheeler. And work completed that gets to be redone because of a change gets billed as an extra.

Consulting engineering firms bill like attorneys: by the hour. If you added up all the hours they billed all their clients for in a day for one of their people that person had to have worked 28 hours that day. They sell this on the fact that they have to pay that person for 8 hours and sometimes don't have enough billing to pay for them. So clients have to pay for that person to be available even if they are not used that day. That and the fact that they have set times for things. A telephone call to a client is a half an hour minimum. If it takes five minutes so be it. Billing rates by hour are, as stated previously, a multiplication factor on the person's salary. Sounds like robbery but a consulting firm's actual ROIC is in the 15% to 20% rate if they are busy and very good. Usually it is less than 15%. Not robbery.

One additional item to consider: Make certain that there is a clear understanding of who "owns" the designs and drawings for which you paid the consulting engineering firm. You need to own them. If you carefully read the boilerplate in a standard American Institute of Architects or other trade association proposed contract format, you will find that there is a clause that says the consultant owns them.

Why? So they insure that you must come back to them for any additional work on the facility. You could pay another firm to develop drawings of the existing facility. But this is expensive, so you go back to the original firm for additional work. Yet you want the option of going to another firm or doing additional work in-house. Insure that the stuff you paid for is yours and so note it in your contract with the consulting firm. Then your competitors will not get your designs – they will get their own even though they will look almost the same.

Using an outside consulting engineering firm is something that most companies end up doing and doing often if they are expanding their market or sales volume. Learn to select good ones, manage them carefully, manage cost very carefully, and insure that the product you get from them is what you need. Over time you will find a few firms that work very well with you and that you can call on in a pinch.

If you have a disaster or crash project it sure is nice to make one phone call and get a group of qualified professionals on the next airplane ready to help you out.

Engineering Management, an Irreverent Primer

Section III, Chapter X

THE ROLE OF PURCHASING

To many engineers, purchasing, that is the separate Purchasing Department, is a barely tolerated nit picking distraction. Purchasing seems to enjoy their position of serving as road block, delay generating unit, or designated cheap skates. In most organizations purchasing is the only entity permitted to actually execute financial commitments for the purchase of goods and services. Since they are not engineers, they usually do not have a clue what they are purchasing. They just know how to do it in a way that protects the business and delivers good value for money.

My opinion is that if there wasn't a separate purchasing group in your organization it might be a good idea to invent one. What? Burn the heretic!

As an engineering manager you need a separate purchasing group because its very existence acts as an important check and balance to your exuberant engineers bound and determined to buy the equipment they want, where they want and to hire contractors who they think execute quality work quickly. And purchasing does not exist to frustrate engineers. It exists to insure that purchases are made based on good quotations, from the best and lowest cost source, in accordance with company policy and the rules of contract law.

All commitments to buy materials and services are legally binding contracts. If you ask a vendor to deliver something and accept the delivery you must pay the vendor. If you do the asking verbally, it is still a contract. If you ask without getting a price and the vendor delivers, you have to pay the vendor whatever he/she invoices you, even if the price is extravagant.

The other alternative is a law suite. Remember, when attorneys are involved, usually the only people who get rich are the attorneys. This is why most insurance companies will settle out of court to stop the legal fee bleeding regardless of the strength of their case. Law suits and counter suits burn money like a bonfire. They tie up your people with seemingly endless interrogatories and dispositions.

If you get into a potential law suit situation with a vendor or contractor, make an attempt to sit down with one of their senior line managers and negotiate. Be flexible and do not get into a confrontation. If your adversary is in touch with their anger and you are also you have an impasse and the attorneys win. Talk turkey. It saves money.

Example: A company hired a consulting engineering firm to do a major ($10MM) utility project. The initial contract was for $1.2MM and the scope involved essentially doing everything – all seven phases from studies through start-up and support. Procurement of equipment and hiring contractors was retained by the company but was to be based on the consultant's recommendations. At 12% of the project the consultant's fee was within the normal range of engineering costs.

The project dragged since it took far longer than anticipated to get the environmental permits. There were changes that environmental agencies insisted upon. When billing reached the $1.2MM, the consulting firm was told that their contract was terminated, they were paid $1MM and the company went on to finish construction and start up the facility themselves using their in house engineering staff. The consulting firm invoiced for another $760K in extras plus the $200K. Purchasing said no. The consulting firm filed suite for the $960K.

In this case it was obvious that the consulting firm thought that it was hired to do the entire job and was to be paid what ever it took to do so. The company thought that they hired an expert who should have known about environmental requirements and so there were no extras – it was a firm lump sum bid. In fact the company felt that the consulting firm should reimburse the company for part of the construction management it paid the consultant for but never received. A detailed legal review of the performance specification and contract indicated that they were vague enough that neither interpretation was solid. Both the company and the engineering firm dug in their heels.

In house legal staffs of both antagonists hired outside counsel with experience in contract litigation. Attorneys for both sides filed suits and counter suits telling their clients that they each had excellent cases. They also started in with page after page of interrogatories and set up hours and hours of dispositions for their own side and the opposition. The legal fee meter was spinning like the dials on a slot machine right after the handle had been pulled.

Finally the company's technical director called the VP of operations for the consultant. Warily he agreed to meet informally at a neutral location but he wanted to bring his attorneys. The technical director said no attorneys – just us. Ultimately the meeting happened is a small room in a local restaurant for lunch.

Each side stated their case. The technical director showed the VP invoices for his side's legal bills to date: $134K. The VP admitted that he had equivalent bills. So combined legal fees were $268K about half way through the law suit process. This was 28% of what the original suit was for and neither side had a strong feeling that they would prevail in the end.

The technical director proposed that he pay the consultant the $200K from the original contract and call it even. The consultant accepted. So when the dust settled the consultant got $1.2MM but after legal fees ended up with $1.066MM. Given the consultant's actual cost, not billing rates, they lost about $300K. The company ended up paying $1.334MM for $1.0MM worth of value received, a loss of $300K. The attorneys made $268K.

The company's purchasing contract along with a bullet proof performance specification could have stopped this from the outset. For the company the project manager was chastised for a poor performance specification and the purchasing agent for a poor contract. On the consultant's side their purchasing group was chastised for poor performance and their project manager got heat for not insuring that what he was doing and what the client expected matched. Everyone lost but the attorneys.

This is typical! Don't let it get to this point if you can help it.

Quotations: It is in the business' best interest to request quotations from multiple vendors. It keeps them all honest and competitive. Although your engineering staff will tell you otherwise, most equipment and material can be specified in a way that it becomes generic. If your engineers are only specifying one manufacturer's equipment it limits the pool of available vendors.

Of course, many businesses have standards establishing one manufacturer for generic items like electric motors, stainless steel pipe fittings, lubrication oil, or even paint. This is done since purchase prices have been negotiated in advance by purchasing and factory stockrooms need only stock spares for one manufacturer's material. These standards should be incorporated in your Engineering Standard Specifications. A change to a standard from one manufacturer's item to another's is a big deal with far reaching financial consequences. It should be done carefully and engineering, and even your divisional manufacturing management, needs to be involved.

Most purchasing departments have lists of "qualified" vendors for standard equipment. An engineer sends a bid package consisting of a request for quotation, number and type, specification, and special criteria to purchasing and they query their listed vendors. The engineer gets bids back and usually is required to select the lowest bidder all things being equal.

If the engineer needs a bid on construction or equipment installation, the same applies but drawings, specifications, finish schedules, material call out lists and such are usually involved. Purchasing should be maintaining a list of local contractors who are skilled in their particular or multiple trades and who have built up a relationship with the business over the years. Specifications and drawings for construction and equipment installation need to be detailed and precise.

Once purchasing has bids back, they can elect to negotiate with a few of the bidders, usually the low bidder and maybe the next lowest. This is a good policy and results in lower prices usually. The operative word is usually. Every bidder has the right to a reasonable profit. If your bidders are cajoled into reducing their profit below say 10% of the bid, they can't sustain their business that way. If your purchasing group has strong armed a struggling contractor into a low ball bid on a construction project, you can bet he/she will cut every corner and bombard you with extras every step between inception and completion – if they even manage to last until completion.

Rest assured that getting another contractor to complete work that a struggling contractor had to walk away from is very, very expensive. They will usually refuse to bid or if forced, will really pad their bids. They all know that they will be taking over responsibility for work that was done on a shoe string, and that they will have to do a lot over.

So if your purchasing group is prone to negotiation after bids, make sure you are aware and have had a manager to manager meeting to set up guidelines for negotiation.

Three bids are essential for insuring that you are getting good bids. Four bids are better. Then you can evaluate the bids easier. And the bid spread tells you about the quality of the drawing and specifications your group submitted with the request for quotation to purchasing. Close bids show a good package. A wide spread shows potential problems. Were the bidders really bidding on the same work as they perceived? My recommendation is that if you have more than a 25%, relatively equally spread difference between three bidders, you have a specifications or drawing problem. Review the package, refine it and requote. Unless, of course one bidder admits to purchasing that they are just too busy and only bid high to insure they have a chance to bid again on another project in the future.

Sometimes your engineering group just has to single source something. What you need is the only one of its kind or what you need has to be fabricated from scratch. Ok. So be it. Get purchasing to call in a couple of qualified vendors and discuss your requirements. Sometimes you will have to go to a trade show to find someone. Take a purchasing agent with you.

When a vendor or contractor starts to try to pull a fast one and create their own extras invoicing you for more than their bid amount, let purchasing handle it. You can tell purchasing to play hard ball. Normally they will without your intervention. Remember, engineering specifies, purchasing buys and administers contracts. It is a neat distinction that keeps you out of financial hassles with vendors or contractors most of the time.

Officially, in most businesses, purchasing approves invoices for delivered and accepted material or services and then accounting (accounts payable) pays the vendor. Purchasing will not approve an invoice on a large item if the appropriate engineer hasn't approved it first. But purchasing can and should take exception to approvals for more than the bid amount without a change order – preferably issued by the engineer in draft and formally by purchasing to the vendor before the change occurred.

In most businesses purchasing has set up blanket orders with local suppliers for material and small equipment that the business purchases on a regular basis. These blanket orders have been bid and negotiated quite significantly and consequently only a release is required to get the material on the way. Insure your engineers understand this. It saves a lot of time.

It should be noted that an experienced capital material and construction services buyer is critical to the success of your organization. It takes years to develop this critical skill and hopefully, your business has at least one.

A new engineering manager should make an effort to sit down with the purchasing manager and discuss purchasing's role. You will undoubtedly be told that purchasing is the most important department in the business. After the shock wears off, establish your own definition of purchasing's role and then split the two roles right down the middle in your mind. It is the checks and balances between engineering and purchasing that are important.

CONTRACTOR/VENDOR RELATIONS

It is in any business's best interest to maintain a list of qualified and capable contractors near each site. A purchasing manager told me once that engineering could disqualify vendors three times as fast as purchasing could find new ones and qualify them. That's probably true. Help your purchasing group find and qualify vendors. Help purchasing maintain at least three in each material or construction skill area.

Understand that your vendors are in business like you are to make money. They provide a service and have a right to a reasonable level of return on their investment. Don't expect them to provide material or services below their cost very often. On a few occasions they may be willing to accept a contract for work on a break even basis. This is usually the case when work in the immediate geographical area is in short supply and the contractor wants to retain his best workers. Always expect them to want to be paid for materials or services they have delivered quickly and without a hassle.

Purchasing should be the primary group in your company that is responsible for contractor and vendor relations. It is not your patch to worry about. That said, it is your responsibility to assist purchasing. If engineering produces good specifications and drawings that accurately and completely depict the work to be done, if engineering has a good set of general and detailed engineering standard specifications, and if vendors are given adequate time to prepare bids, you will get better value for money.

A good vendor or contractor will ask questions. They will propose alternatives that may help your engineering group execute faster or save money. Some engineers get upset if a contractor questions their design. This is ego, not engineering. A good contractor has years of experience installing equipment and has a good understanding of how, and in what sequence, installation seems to proceed the smoothest and cheapest. Good ones will share this knowledge with you.

Unless you have been lucky enough to have actually worked as a millwright or electrician, etc. you may not know the methods employed to execute work. A four inch conduit with three each 250 mcm wires in it may be just a line on a drawing and was selected by consulting a table in a code book after some calculations. But each wire is the size of your thumb, is heavy and the conduit is larger in diameter than your coffee cup. It doesn't bend or flex much and it requires a lot of knowhow and muscle power to install an electrical feeder like this.

After about three years in engineering and watching contractors pour and finish concrete, I assumed that it wasn't difficult and something I could do myself. I can remember calculating that I needed two yards of pea gravel concrete to put a four inch topping on my old home's basement floor. So I took a days vacation and had a concrete supplier show up with a ready mix truck at my home early one morning, sticking the dump chute through my basement window with me below it with a wheelbarrow. Conceptually, two yards of concrete is not a lot. In reality it was an enormous amount, and even though I kept the truck there for hours parceling out one wheelbarrow at a time, it was 20 hours of continuous backbreaking work later before I had the concrete spread and evenly remotely level. The finish looked like an old cobblestone road. I have never tried to place and finish concrete myself since.

Unless you are intimately familiar with the techniques contractors use to fabricate or install materials, leave the details of how they actually do it to them. What you want is performance and they know how to provide it. You don't.

As noted previously, change orders are the single most problematic area. And there will be change orders in most projects. If they are handled carefully, both engineering and the contractor or vendor will remain satisfied that they received value for money.

Change orders need to be handled with some formality. An engineer, or any representative of your company telling the contractor's on site foreman to do some additional work verbally is, legally, a change order. Contractors need to understand who is authorized to speak for the company and who is not. Include this information in the purchase order or construction contract.

Then, when a change is needed an authorized representative can meet with the contractor's foreman who can then call his office and have them come up with a quick estimate of the cost of the change. Good vendors or contractors will get back to purchasing with a revised cost quickly. Usually this is in a not to exceed format, i.e.: "we will do the additional work on a time and material basis for a not to exceed price of $XXX."

Any time and material work, including "not to exceed", requires the contractor to submit a detailed accounting of what they spent in material (copies of their invoices) and labor (copies of their time sheets). To this, normally, they are allowed to add 15% for overhead and 15% for profit. If they gave a not to exceed price, then that price is the upper limit including overhead and profit. Purchasing cuts a T&M (NTE) change order and the contractor executes the work. The contractor invoices for the work, with documentation, and the engineer and purchasing authorize payment.

It is absolutely essential that the engineer keep very accurate track of all change orders issued either verbally or in writing. There are engineers who have lost their jobs over issuing verbal change orders willy nilly with out recording them or advising purchasing. When the invoices for this work arrive, invariably they are far more expensive than the engineer thought they would be and his or her budget may not even be able to cover them. He or she then gets into a pissing session with the contractor as he or she tries to wiggle out of a self-inflicted crunch. The contractor has been ill served, purchasing has been kept out of the loop and all are unhappy.

There are a few unscrupulous vendors and contractors. Some believe that they can feather their nests by providing gifts to an engineer or purchasing agent expecting special favors later. Accepting gifts from vendors or contractors is bribery on the part of the contractor or vendor and embezzlement on the part of the engineer or purchasing agent and can not be tolerated. Some businesses notify all vendors and contractors that it is against company policy for any representative of the company to accept any gift what so ever. Some draw the line at maybe a bottle of booze at Christmas, providing it is given to all who have interacted with those particular company representatives over the year.

However, sports tickets, fancy dinners out, small electronics, cases of wine, work on a representative's home, etc. are out. Make it a policy that if a vendor or contractor buys dinner this time, the next time your company's representative buys. Deal with the dinner checks like you would with your friends – go Dutch or take turns picking up the checks.

If your business follows this policy you will find that your vendors and contractors will respect it and will feel that the best way to stay in your good graces is to do a good job delivering on their commitments. Your employees who violate this policy should be subject to termination.

Vendors and contractors who cut corners, slip in materials that are below standard, try to low ball their bids and then make it up with extras, double invoice, or other unsavory practices should be summarily removed from your company's authorized bidders list. You have to send a strong message from moment one – ethical business practices and value for money are the only things that should define your vendor and contractor relations.

UNION VS NON-UNION

There is a special situation that some companies get themselves into. That is dealing with the difference between union contractors and merit shop (non-union) contractors. In many cases a union shop will not work on the same site or especially the same job with non-union contractors. There are exceptions to this. Some union shops don't mind working along side an owner/operator: someone who is providing a service him or herself as the proprietor of his/her own small business.

In many areas of the country, and especially if your site has a union, you are restricted to using contractors who employ union building trade workers to execute the work. Invariably, union contractors are more expensive. And surprisingly not because they pay their workers significantly more. Skilled workers usually get paid quite well regardless of their union status. But the overhead for union shops is higher – either for union dues and contributions to union pension funds or medical plans, or because there are relatively restricted work rules like how many union foremen are required or how many apprentices are required for each job based on the total trade manpower on site.

Union shop contractors have a better source of skilled workers. They can go to the union hall and request them. Merit shop contractors have to go outside and hire skilled workers. And you may find that there are contractors who burn the candle at both ends. They have a union shop division and a merit shop division. Some workers will work on a merit shop job if there is a shortage of union jobs. Although against union rules, some unions look the other way if jobs are scarce.

Building trades unions are broken down by trades. There will be iron workers, pipe fitters, electricians, plasterers, painters, insulation workers, laborers, heavy equipment operators, masonry (brick layers), cement finishers, millwrights (carpenters) and etc. All have a pretty good idea what is their work and what belongs to other trades and these overlap. Then there is a jurisdictional dispute that stops work. It is frustrating.

And some unions are far more militant than others. In each geographical area there should be one individual that is the president of the local Building Trades Council: the one person who has been elected by the unions themselves to be the contact point for all unions and union contractors to help resolve disputes. If your site is in a union area, try to meet (you and purchasing) and cultivate a relationship with this individual. You will find that this individual can help. Especially if a union from another area is causing problems at a job site.

You should note that if a union sets up an informational picket line due to a jurisdictional dispute, in most cases, other union workers will not cross it. Even truckers (teamster's union) will not deliver to the site. Work stops. Worse case, and if you do not have a separate designated site entrance for construction, an informational picket can stop all deliveries including raw materials in and finished goods out at a factory. Always have a separately identified and posted site entrance for construction. It is illegal to picket anywhere but at a construction site entrance in most states. The pickets may shut down the construction job until the dispute is resolved, but at least they won't shut down your adjacent factory.

Your purchasing group will have to get on the contractors, the president of the building trades counsel, and the individual union business agents to resolve the dispute. Threatening to shut the job down and throw all contractors off site is a good ploy if you really can do so. In most cases negotiation is essential.

There are cases where a few bad apples take action to actually destroy work in progress that has been done by another trade. They break things, smash or damage materials and destroy property. I have witnessed a situation between a carpenter's union and a plaster's union where both claimed the work of hanging drywall. The work was given to the carpenters by the contractor and a few plasterers took it upon themselves to kick holes in the installed drywall.

This is criminal behavior and can not be tolerated. If you can determine the culprits, call the local police and prosecute the perpetrators to the full extent of the law. If you can't, but know what contractor they were working for, haul the contractor in and tell him or her that it is his or her responsibility to control their workers, and that he or she needs to not only do so but pay for the damages. If he/she refuses, unilaterally void the contract and throw the offending contractor off site. You need to establish a zero tolerance for vandalism.

Some economically disadvantaged areas have recently seen a trend where union shops are being replaced with merit shops. The unions have basically priced themselves out of the market. There is some turmoil in these regions but ultimately, until the local economy rebounds, if it ever rebounds, the trend continues.

My personal experience says that the quality of work performed makes little difference: union or merit shop. It is true that unions have some strict rules for how much on the job and even formal training and experience a member needs to obtain prior to moving from apprentice to journeyman and ultimately to master. So theoretically, a union provides workers with a higher skill level. True in heavily unionized areas since most jobs are union. Not so true in lightly unionized areas.

If you look hard enough you can always find merit shop contractors with experienced and capable people, even if you must import them from outside the immediate site area. If there are union trades sitting on the bench at the union hall in your area at the same time you are importing non union men from outside, expect grief. The union membership will demand that the union take some action. And that action may well be a picket line. Non union workers will attempt to cross the line and you could have an incident. If you elect this route insure that your local police are on site to preclude violence. It may be true that hell hath no fury like a woman scorned but taking jobs away from union members in their own back yard with non union people from outside their area comes in a close second in the fury department.

Section III, Chapter XI

FAST TRACK

The business always wants engineering projects to be on a fast track. But there is a technique for really moving a project rapidly by taking the normal project sequence of design, quote, contract and construct and short circuiting it. It is called "fast track" and should be called "risky and expensive fast track" to insure all management understands what is to be done.

As discussed in the Management Principles in the front of this book, you are trying to execute a project fast and with precision. And it will cost a bundle.

Every project has what is perceived as a dead band of time when, to the outside observer, nothing is happening. The engineers are designing the systems and developing specifications and installation plans. They are getting quotes on equipment. And this engineering takes time. So a lot is happening - it just isn't visible.

A normal project proceeds in sequence. Preliminary engineering, detailed or final engineering, quotation and equipment purchase, quotation and construction contracts, buying material and construction start, construction completion and start up.

What is done on a fast track project is to short circuit the sequence. That is, to quote materials and construction based on preliminary drawings and specifications. Once the quotes based on the preliminary engineering come in, the business purchases these materials and services and in some cases even starts construction before the final engineering is completed. I have seen building foundations being poured before the wall structure was selected. So they were poured for the worst case heaviest wall load. This is typical and fast track doesn't leave time for much value engineering. Pick the largest item you think you might need, worst case, and buy or build it.

You then handle the differences between preliminary and final engineering with change orders. Lots of change orders.

To say that coordination between engineering disciplines and between engineering and purchasing is critical in a fast track project is an understatement. Someone has to insure that all the designers and engineers are marching to the same drum beat. Otherwise you will have a lot of wasted time redoing things, and the actual construction or equipment installation will take far longer than anticipated, and cost a fortune as things are installed, torn out, and reinstalled.

If you can talk your upper management out of fast track, do so. If not, make sure they understand the financial risks. You will be bidding and constructing with incomplete data, and there will be conflicts and changes required on the fly. A fast track project can be successful, but be prepared to spend 15% or more above a similar "normal" project

There are times when getting a project up and running early has such a significant financial incentive that the additional capital cost is more than paid for. The ultimate fast track project is a disaster of some kind. Here time is critical and the business has to get the facility back on line as soon as possible. Then engineering becomes almost strictly arm waving in the field and teams of contractors move quickly to execute the engineer's verbal orders. A good purchasing agent should be on site to coordinate that aspect and keep score. Only your most experienced engineers should be called on to manage a disaster project – or maybe you better manage it yourself.

One thing that really helps in a disaster project is a detailed set of as built drawings on file from the most recent expansions or alterations. A disaster like a fire or explosion doesn't leave much to start with and having drawings helps you understand what was there previously.

Most engineering organizations are tasked with the responsibility of maintaining an accurate depiction of all the fixed assets owned by the company. So all projects should have as built drawings on file. In today's world these are usually computer aided design files on DVD ROMs at some off site secure storage facility. Older ones will be on microfiche or even paper. As builts are not only essential for disaster projects but for any project especially on an older site. Carefully document any underground utility locations when they are installed. Finding a water main or an electrical power feeder when a back hoe pokes a hole in it is not fun.

Engineering Management, an Irreverent Primer

Section III, Chapter XII

GREENFIELD SITE

This is the project every project engineer dreams of: a chance to start from absolute scratch and design and build a stand alone facility. It is an opportunity to "do it your way," and to show how much you have learned living with the mistakes of others in the past as you expanded or modified facilities originally built by them.

Brace yourself. It is not all fun. There are pitfalls. Was the site previously occupied by something besides nature? If so there will be remnants. Usually underground remnants. I have seen one site that previously was occupied by a small business that used a coal fired boiler. The old structure and boiler were gone but the previous factory did not haul their coal ash off site. They buried it. Neat. You can not run steel fire main pipe through buried coal ash. When the ash gets wet it becomes corrosive – very corrosive. Your fire main could easily become a very expensive underground irrigation system.

How about underground fuel oil tanks? If not maintained they ultimately leak. And you are faced with a pollution site to clean up, and a lot of fun supervision from the local environmental protection agency, and lots of fun tests to pay for, and reports to write. Even worse is a site that pumped toxic waste underground or buried drums of it in a hole. Underground plumes from toxic waste travel long distances and although your company is not directly responsible, they will be considered part of the solution. The best you can hope for is a natural undeveloped site where the previous occupants were not human.

Site selection. Choose your site carefully. Most businesses have carefully prepared site selection criteria. Raw material sourcing and transportation to the site, finished goods transportation off the site, and energy availability are important. How about waste water treatment if your industry creates waste water? If not, how about a municipal waste treatment plant for sanitary waste? Real estate taxes or tax breaks? Is it in the hundred year flood plain? Subject to storms or storm surge? Availability of skilled labor, contract labor?

Then you must take a hard look at the site itself. Just how much site clearing and overburden relocation will be involved? What is the soil's bearing ratio? Is it rock you will have to blast or sand so that you will have to shore every penetration or put in lots of sheet piles to preclude cave inns? Are you going to have to drive piles for foundations? Potential for sink holes? Is there any quick sand or plastic clay? Does it drain well in a major rain storm? If you dig a basement in an area with clay soil and then backfill outside the foundation and wall with gravel you have essentially built a swimming pool. Heavy rains will fill the clay hole and basement unless you install lots of French drains and sump pumps.

Is the site large enough to create a buffer between you and your neighbors; enough for the future when you have to expand the factory? What are the zoning and set back restrictions? Do you have an up to date survey?

I could go on and on. A perfect site is one that is cleared, flat, well drained, has good road-way access and utility mains nearby, is close to labor, and close to either your market or your raw materials. Usually these sites are in an industrial park created by the local government to attract industry. A site in an industrial park may be more expensive initially but may well be cheaper in the long run. This, and government tax incentives, is why many companies choose industrial park sites for their new facilities.

If you get an opportunity to have input, an industrial park is a good bet. Once you have a site in mind, go visit your potential future neighbors. Talk with their engineers and if they agree their architect and general contractor. You will get a lot of insight as to what to expect once you break ground for construction.

Once a site has been selected and purchased, it is site layout time. This may well require a topographical survey and soil core samples. Layout your entire site, boundary to boundary including your proposed factory, offices, roadways, parking areas, shipping and receiving areas, areas identified for factory expansion, areas for utility systems and waste handling, and areas held in reserve.

Then, and only then, are you prepared to do a detailed building footprint layout. If you do not have the required in house staff of architectural and civil engineers, and very few business do, you need an architectural firm, an industrial architectural firm. Many businesses will want an aesthetically pleasing structure. Yet some will elect a utilitarian structure. You need to select an architectural firm that is willing to meet your needs not their own to create an edifice.

The selection process is very close to the consulting engineering firm selection process noted previously. Go look at completed equivalent structures the architectural firm has designed. Talk to the owner's representatives. Were they happy with the service they got from the architectural firm?

We had a joke about site selection in one firm that I worked for. The criteria were as follows:

Fly into the smallest airport that can accept a commercial prop jet aircraft. Rent a car and tune the radio to the loudest FM radio station. Drive whatever direction is least developed. When the radio station fades start looking for land to buy.

Of course that was not true, but we did have some sites that were really a long way away from nice hotels, restaurants, suppliers and contractors. The land was cheap.

BUILDING TYPES

Once you have selected a firm, consider building types. Although initial cost is important, life cycle cost is more important. A structure that was cheap initially will bleed your company for years with high maintenance costs. A facility that is difficult to keep clean will look shoddy and unkempt the moment a squeeze is placed on the janitorial budget.

Think about the facility structures your company owns currently. What are their good points, their problem areas? What would you do differently if it was your money?

Slab on grade (with grade foundation beams), structural steel frame and tilt up precast concrete walls, and a light weight precast concrete roof deck are typical utilitarian industrial structures in the developed world. Save the fancy architecture for the office building out front.

What you are looking for in any factory is a clear span open space structure with minimum columns that can be easily maintained and cheaply expandable. Structural steel can facilitate huge clear spans, although much over sixty feet starts to get pricy. A structural steel frame can be built to handle overhead cranes. Roof decks should slope to drains. Dead flat un-drained roofs suffer from rapid membrane failure.

If you're just looking for a small structure, consider the pre-engineered metal buildings. Just mount them on a four foot or higher poured concrete wall foundation. They will last a lot longer and any structure that is going to be using fork lift trucks internally needs fork proof walls and column protection or the fork trucks will eat holes in your walls and knock columns off their foundation in short order. Fork lifts do an enormous amount of damage to a structure even with driver certification and training programs. The forks are spears and the counter weight is a battering ram.

Masonry is good for applications without fork lift trucks. It is inexpensive in the concrete block format. There will be pilasters every so often to hold the wall up. You can even get block with an exterior architectural texture to please the eye. If you are going to use masonry with fork trucks, you need a huge steel angle solidly lagged to the floor all around the block interior perimeter and large steel pipe backfilled with concrete to encase structural support columns.

Floors are the single most vexing problem in most process industry facilities. Anywhere you are going to have water as part of the process or for cleaning will cause floors to fail. Anywhere you will also have toxic or corrosive liquid your floor is in jeopardy. The best anti corrosive floor is acid proof brick set in furan resin over a well sloped concrete sub floor with lots of catch drains. It is very expensive. The next would be one of the newer epoxy floors, again well sloped and with catch basin drains. Urethane coated concrete is not for process areas in any industry where water or corrosives will fall on the floor. Acids and caustic materials will eat the urethane off coated concrete which will then spall and fail very rapidly under those conditions.

For metal fabrication, foundry and mill facilities there is nothing quite as good as a wood block floor. It is resilient, can take a huge amount of dings and wear, is easy on anything dropped on it and will last for years. It also is expensive, and when worn, almost impossible to keep clean.

Terrazzo is great for light industries. But stained hardened concrete with a good grade of polyurethane is cheaper and almost as good. Epoxy terrazzo is also good but expensive and difficult to place smoothly.

Then there are the ceramic tiles. It should be noted that there is a product called cannery tile, 4" x 4" x 1" thick baked and glazed clay. It is misnamed. It should never be used in place of acid proof brick. It cracks and then breaks from the smallest ding, gets water under it and comes up. Otherwise, ceramic tile is great if you are willing to clean the grout lines and periodically re-grout. Remember that any hard floor, and ceramic tile is one of the hardest, is tough on workers feet.

Asbestos tile is out for obvious reasons. Vinyl tile shrinks and expands significantly with temperature variations. Sheet vinyl does also and develops lumps or ripples over large areas. In offices there is industrial grade carpet, and even some of the newer thin prefinished hardwoods. Laminate has such a thin surface layer that it will scratch and deteriorate in any heavy traffic area. Prefinished hardwood can at least be sanded and refinished if necessary. A good office compromise is terrazzo in the high traffic aisles and low nap closed loop industrial carpet (metal edges) in the other areas.

Areas with a lot of computer cables require raised floors. Make sure you protect the area under the raised floor with a fire suppression system. Wifi works well in an office environment if, and only if, it is secure and I mean totally secure. Password access yes, but a positive reverse ping to a registered computer is also required. Consider encryption. Remember that anytime you transmit voice or data over the airwaves you have a risk of interception by unauthorized hackers. This includes Wifi but also cellular PDAs, Bluetooth, VHF voice, etc. Any transmission, even very low power, is subject to interception.

Wifi doesn't work well in factories where there may be a lot of machinery and associated electromagnetic interference. Factory areas need a communication backbone. Here fiber optics or shielded hardwiring is required. Run the backbone in conduit and insure there are connection nodes in each area.

Square floor-wall junctures are crap catchers. As are horizontal structural frame members. Try to specify cove base corners and minimize stringers and other items that will catch and retain dirt. Open base equipment mounts and frames catch and retain dirt rapidly. If it is even a little inaccessible it will not get cleaned. Out of sight out of mind. This is critical in a food or pharmaceutical facility but lessons learned there are applicable to all facilities. Clean design facilitates keeping a factory clean.

Each business should have a set of guidelines they have developed over the years related to the structural details they want in their buildings that minimize life cycle maintenance cost and allow their facilities to remain clean and appealing. Even a steel mill or foundry can be kept relatively clean if properly designed. And cleanliness helps with worker attitude, promotes safety, and presents a good image to anyone touring the facility.

Section IV

SPECIAL ENGINEERING SITUATIONS

Section IV, Chapter I

DEVELOPING WORLD

There is an old adage that if you think you know how to build a factory, try it in a developing country. So much of what we take for granted in the developed world just does not exist there. There are few large local plumbing or electrical supply firms, maybe no ready mix concrete firm, and materials you may be used to using in the USA must be imported at significant shipping and duty costs.

In Brazil, anywhere outside the Rio – Sao Paulo corridor is lacking in material availability. I once spent half a day in Recife looking for a one inch brass gate valve – something any hardware store in the USA would have lots of. Of course the local hardware store managers all told me they could get one trucked or flown in from Sao Paulo. Only took two days to fly it for five times the cost of the valve. A truck took two weeks and cost twice the cost of the valve.

And if material is available, it usually is not even close to the quality you are used to. Don't be surprised if the government will not permit the importation of material they believe is available locally or somewhere within their borders – even if what is available is of very poor quality.

In general, labor is cheap and fancy equipment very expensive. Their method for excavating a twenty foot by twenty foot by four foot deep foundation hole involves thirty men with picks, shovels and wheelbarrows, not a back hoe and a truck. Cutting an eight inch hole in a concrete floor involves one man with a cold chisel and hammer and three days of work instead of a core drill and an hour. A hand laid cobblestone driveway may be cheaper than a poured concrete one and a better bet since the quality of concrete is suspect.

Coastal areas may be somewhat developed but interiors are invariably agricultural and poor. If your time in country is spent in the major cities or affluent suburbs, you may not see the poverty, but it is there. And it drives the culture.

Much of the construction labor you get comes from the interior, can be illiterate, sometimes unskilled and you or your contractor must provide temporary living quarters, including sleeping, eating and sanitary facilities on site or near by. Then the workers will work three weeks straight and then take four or five days to go home for a visit.

Any tool, especially small power tools, you import grows legs almost immediately. A quarter inch electrical drill is worth two weeks salary to a construction worker. And in most locations a black market in pilfered tools and equipment flourishes. Anything not nailed down disappears. And they will pry up what is nailed down if they have the time. Site security is essential. The first order of business should be a fence, gate, gate guard and roving patrol. Every person or vehicle departing the site should be subject to search.

We kept an opaque jug with different colored marbles in it. As you exited you poured a marble out. If it was black your car, briefcase, lunch box, or parcels were searched by the on duty guards. If it was white you got to go without a search. The rule applied to everybody: management, hourly, contractors, vendors – everybody. If you wanted to tighten security you just changed the ratio of black to white marbles in the jug.

The developing world is plagued with a poverty problem. Roughly five percent of the population is rich, runs the country and lives as we do or even better than we do. There is a very small middle class that gets by with less than what our lower middle class survives on but considers themselves to be well off. Then there are the masses of poor that live in shacks, participate little in the economy, can't afford to educate their children, have little access to fresh water and no access to sanitary facilities, and live in quiet desperation.

To say that crime is rampant would be an understatement in many countries. If it isn't, it is because the ruling class maintains a police force that has been told to be absolutely ruthless with small time criminals. Shoot first and ask questions afterwards is practiced as a rule. If there are locally grown drugs available, there is a thriving and lucrative smuggling business, and the police can be on the take. In many poor areas the drug lords are the defacto mayors and run the area with an iron fist. The police don't go into these areas unless essential and then only in force with armored vehicles and automatic weapons.

However, the majority of the poor are in survival mode and quietly suffer along hoping for a better day tomorrow. They eat rice and beans and pray for a chicken. They live in "ranchos" or "favelas" in huts perched on hillsides. They climb telephone poles and steal electricity to run their one or two light bulbs and invariably a small black and white TV. Some get fried to a crisp trying to hook their wire onto a live transformer. They barter for what they can't buy or steal. And they have babies.

You will find that the life span of the poor is quite short compared to developed world standards. Medical care is almost non existent or provided by understaffed and crowded medical clinics. So most countries have a very young population. And this young population has a strong desire for anything American or European. This demand is what drives developed world countries to set up production facilities in the developing country.

There are the cultural differences to contend with. The pace of life just isn't the same as it is in our fast paced society. Things happen slower. Time is not of the essence. A local may well feel that if he or she shows up for a meeting two hours late that they are on time. Business meetings start with a half hour or more of social discussions prior to getting down to the task at hand. If someone tells you that they will deliver something on Monday and it shows up the following Friday, they think they were responsive and punctual. And the locals will not understand or be insulted if you demand a faster response.

Social norms are different and do not necessarily follow our puritan morale standards. They look at sex differently. It is just a part of life and premarital or extramarital sex is not a big deal to some. Prostitution may be legal and accepted, or at least tolerated, as a way for poor women to survive. There are professionals, semi professionals and enthusiastic amateurs augmenting their normal income. In some areas women outnumber men since the men have gone off to the big cities for work. And a woman who has a child out of wedlock is pretty much screwed. There are few child support laws, and if there are any, they are not enforced.

Of course everyone expects there to be sexually transmitted diseases in the developing world. Although not publicized, the rate of infection can be significant. Consequently most educated locals are very careful and practice safe sex.

Don't be fooled by the fact that a country is listed as being say 95% Catholic and expect the population to follow strict Catholic norms as taught in the USA. They may be Catholic but it is their form of Catholicism. Violating church teachings all week and going to mass on Sunday is considered break even and acceptable.

The rich live well. Income taxes for them are either low or not enforced. They have nice houses, usually surrounded by walls with gated entrances. All have household help: maids, gardeners, drivers, bodyguards, and home and office security guards. Many of the rich carry concealed hand guns.

Depending on the number of poor and how much the rich allow to trickle down there are opposition political parties espousing some form of socialism or communism as an answer to the poor's problems. These groups can get elected (Venezuela) or, in frustration, set up insurgencies. Other insurgencies are created to deal with ethnic or religious differences. In many cases the borders of developing countries were created by colonial powers without regard to the ethnic or religious backgrounds of those caught inside a border. Strife is almost guaranteed as the discriminated against have-nots struggle against the haves.

An engineering manager sent to build a factory in a developing country has to cope with all the above along with the normal difficulties associated with the job. Added to that is normally the fact that he/she doesn't speak the language. Don't think that if you studied say Spanish in college that you speak their Spanish. It is very different, not only in dialect but in idioms. It takes years to be even remotely bilingual. One expat American showed me a book: *Ten thousand Portuguese Irregular Verbs*. He said that he was told by a local that there were more than that but the book only listed the most important.

You will, after some time there and maybe with some language courses, get to the point where you understand at least some of what is being said. Bide your time. Don't let on. You will learn a lot – some things you may not like – but all are important to improving your effectiveness.

I didn't bother to deal with verb declinations and endings. I just used infinitives and got lots of laughs – especially if I screwed up a gender. But between the local's rudimentary English and my rudimentary fumbling in their native language, I could communicate. They respected me for trying.

Even so I was stumped many times. Literal translation made no sense. It was because I didn't know a cultural joke or story that was common knowledge among the locals. Over time I learned a few but there were thousands and the punch line to these pervaded the conversation.

Governments, laws, and regulations in developing countries are difficult to deal with. Although they are in desperate need for the jobs and benefits to their economy your project will provide, they also look at you as a carpet bagger. And the more you look down your nose at them and the more you measure their country using North American or European standards, the more they will dislike you. But they will not confront often. They will work in the background to limit your success.

Most developing world countries have a system of graduated duties on imported material and articles. The idea is to protect their own developing industries. Since their developing industries don't have the economies of scale that our businesses have, you will find that things like imported automobiles have duties in the 100% range. And the price of any locally produced cars will be 99% of the price of a duty paid imported car.

There was a joke in Venezuela that the number on the back of an imported Mercedes Benz was the price in Bolivars. At the then four to one exchange rate that meant that an E320 cost US$80,000. About $40K more than in the USA then. So the poor ride buses, motor cycles, bikes or walk and only the rich and middle class can afford cars.

Getting an importation license can be a challenge. I can remember attempting to import some US made acid proof brick in one country. I went to the office of the appropriate government official with a sample. I went through an entire presentation on why acid proof brick was critical to maintaining factory sanitation with an interpreter. The official pulled what looked like a lumpy grey cobble stone with a thin glaze on top out of his desk drawer, sat it alongside my perfectly rectangular red acid poof brick and said "Same, no importation license." My guess was that the official's brother in law or something was in the glazed cobblestone business. Ultimately I ended up using a locally available epoxy floor topping.

Driving a rental car in the developing world is a challenge. Traffic laws are generally ignored and unenforced. A roadway will fit as many lanes of traffic as can squeeze in and in either direction. Broken down trucks just stop in the middle for repairs. After dark, red lights are mostly ignored and stop signs are always ignored. Construction barriers are usually just piles of dirt and unlighted at night. The road surface is rough and poorly maintained. Road drainage in a rainstorm may not be adequate and drain grates may be missing. These concrete holes in a roadway can do a number on a tire. Make sure your rental car has a good spare. You will need it.

This would make one think that all these hazards would cause the locals to drive slow. Wrong. They drive like they are on the track at Daytona.

You can take taxis. They are not inexpensive but at least you don't have to drive. If you are not at least somewhat familiar with driving on the left in countries like India, use a taxi. Even so, be prepared for some hair raising experiences.

Once I was to fly from Pune to Mumbai to catch an overseas flight home. I had been in India for three weeks and it seemed like three months and I was more than ready to leave. But it was rainy season and the local Pune airport was closed due to storms.

I elected to take a taxi the 124 kilometers from Pune airport to Mumbai International Airport, nominally a three hour trip. It was raining heavily and the wind was strong. The rains had washed many of the roads out and there were trucks backed up for miles. India drives on the left like the UK. The taxi driver passed these trucks by going fast down the oncoming lane and ducking onto the right berm when another vehicle approached from ahead.

Dodging the jammed main road we went zooming down narrow country roads. We came to what I thought was a lake but he said was a flooded river with white caps on the water surface. He raced across a rickety old bridge stopping to allow the water from a wave that broke over the bridge to dissipate. Then we zoomed forward to stop and wait for the next wave. I could see visions of the bridge collapsing or a wave washing us off it and us bobbing down the river as the taxi sank.

Later he hit a huge pothole which caused all electrical power in the taxi to go out. Bad news since it was night and there we were stopped in the middle of the oncoming lane. He yanked wires out from the radio under the dash, stripped them and wrapped the bare copper around the blown fuse. Inserting it in the fuse holder, we were back in business and off we went again at speed down the right lane.

He knocked the right side view mirror off passing a stopped truck on the left berm. When we got to the road that wound down the face of the Gat Mountains, he dodged big waterfalls coming down the mountain and splashing on the roadway. We fishtailed down the narrow road, mountain on one side and sheer cliff without guard rails on the other. I just held on in the back seat and said every prayer I had ever learned. Then I resorted to beseeching any deity that might be listening and making elaborate promises if I was just permitted to live.

We arrived at the airport seven hours after departure and I paid the driver the $150 we had agreed on. It would cost him more than that to fix what was broken on the trip. Now he had to drive back to Pune. I had missed my flight and would wait around five hours for another flight on another airline.

You may also find that many developing countries have an inflation rate that would bring down the government in the developed world. Prior to 1995

Brazil had a rate approaching 1000%. They recalled the money (back then it was the Cruzerio) and stamped another three zeros on each bill. When you purchased anything on a purchase order it was either priced in US dollars or at the exchange rate in effect the date of delivery. With high inflation rates there is no such thing as home mortgages. Locals have to save up enough to purchase a home or apartment.

Most countries do not allow the importation of used equipment. They suffered from years of North American or European industries importing their cast off old equipment and systems. This perpetuates their competitive disadvantage in the world marketplace. Most countries price used equipment at the equivalent cost for new and then add the appropriate duties. They want the business but they want new modern factories, not something out of the 1890s.

And having suffered from enormous pollution from foreign owned firms in the past, they are leery and insist on good pollution control measures. This is especially true of countries that have recently moved from the third world to the second. I have seen a square mile of light brown foam three feet thick on Lake Valencia in Venezuela as recently as 1983 coming from the discharge of foreign owned meat processing plants.

Normally bureaucracy abounds. Getting all the required permits to construct a facility is about as difficult as getting permits to construct a nuclear power plant in the USA. And rules for permits conflict between government agencies. You need a local agent who understands the system to get through it.

In many countries there is no way you will ever get a permit if you are not using a local architectural or engineering firm to "design" your factory. The government will not even look at your drawings without a local registered architect or engineer's signature. You can get one but they are not cheap. And you will need someone to convert any drawing you have made using English measurement to metric anyway.

A comment about metrification: The whole world outside the USA thinks in meters, kilograms, hectares, Centigrade and etc. Personally, I believe the USA should convert also, but I don't get to make that choice. Learn to convert in your head until you are fluent in metric.

Although the laws in the USA preclude bribes, they are a way of life in many countries. In most countries this must be handled very carefully. Although many government officials participate in this form of additional compensation, those that are dumb enough to get caught are ostracized or even persecuted. Get creative. Pay your local agent and don't ask where all the money went or be prepared to wait forever for the necessary permits.

How about sending a government official to the USA to review one of your sites? Then make sure he/she is appropriately wined and dined while there. Invite an official to inspect your site in country. He/she then signs out of their office for two or three days but spends an hour at your site "inspecting" over coffee and pastry. If someone telephones for them make sure your local secretary has a number she can call with the message.

Another way to handle the bureaucracy is through a local university. They get a lot of good press and mileage out of consulting with a developed world firm and their professors usually are either related or know some government officials. You are going to have to hire some local engineers for your factory anyway. Why not hire them early, or why not hire some who are in their last year, as co-ops? You get them cheap and you get to evaluate the best for future full time hires.

In smaller developing world countries the government and the rich are one and the same and go back generations. Many officials are related. Family names define power. Consider hiring someone with a moderately well known family name, assign him to public relations duties and give him a title like "President." In most countries it is the Managing Director or General Manager that runs a company anyway and your "President" will give you an in with the ruling powers.

Play the game. Invite local and regional officials to a ground breaking or opening, complete with a plant tour and off site lunch at a nice restaurant. Host a dinner party for local officials to meet and greet. Build a day care facility on or near your site and invite the local political leaders and press to its opening. You may need a day care anyway.

Some countries have odd laws and you have to live within them. Venezuela required an on site medical clinic staffed with a nurse and a part time doctor. It was useable by workers and their immediate families and the company paid all costs. There was no requirement to pay an employee's medical insurance. Indonesia required a separate room for prayers complete with rugs and a part time cleric, and workers could use it five times a day for ten minutes each time. Argentina and Brazil required company provided bus transportation from the center of the nearest town if the site was even slightly remote.

Almost all required some form of on site cafeteria for workers to eat (hot meal in the middle of a shift). Food and service were paid by the company, unless the site was next to an area with food service – and then employees had to be allowed one hour for lunch. If you have a continuous process, you sure don't want to shutdown three times a day, so a cafeteria is the best option.

Many countries have strict anti-layoff policies. Usually these involved a form of unemployment compensation where the foreign firm had to provide over 50% of income for a year. But if an employee was terminated for cause a foreign company had to give them six months pay as severance. Not much difference.

You may find requirements that we would call "featherbedding". The government will have regulations like each truck must have two drivers for any trip more than 20 kilometers. All skilled workers must have a full time apprentice. Overtime is limited to no more than four hours per week. Operators can not clean and cleaners can not operate. There can be no more than 20 workers per supervisor. There must be a full time nurse's aid present anytime any worker is working in the plant. Many of these requirements were designed to pad the workforce and increase employment. Actual hourly wage rates might be low but labor costs could be inflated significantly.

I can remember the Venezuelan government declaring that all automatic elevators were not automatic. So every elevator, and Caracas was chock full of high rise buildings due to the shortage of buildable land, had to have an operator sitting on a stool to push the elevator buttons whenever the elevator was in use. Only the designated operators were permitted to push buttons. Apartment buildings, and there were thousands of them, required elevator operators 24/7. It did wonders for reducing the low skilled worker unemployment rate in the capital district.

Most developing countries have some form of profit removal restriction. One allowed a foreign company to repatriate no more than ten percent of the government identified invested capital each year. The important aspect of this was government identified. The government excluded working capital and restricted what they identified as investment to assets purchased inside the country – excluding imported equipment if in their estimation the equipment could have been sourced inside the country. So repatriation really was closer to 5%. Any additional profit went into limbo. The company could spend it in country but couldn't take it out. In some cases a portion of limbo money spent on factory improvement could be added to the government identified invested capital.

Some countries like India only allowed a foreign company to own 49% of a company doing business in India. Of course this was regardless of their investment. Usually companies wanting to do business in India found some local investors to own, at least officially, 51%. Then profit repatriation was limited to a percentage of the 49%. So more was trapped in limbo.

Many countries do not permit foreign companies to own land. They may lease it for a long term (20 years even up to 100 years) but can't own it. At the end of the lease it can revert to the government and all assets on the land revert with the land unless the lease is extended.

The capital repatriation, ownership percentage and land title item noted above change. Some countries seem to alternate ruling parties at each election. So the free traders get in and foreign investment gets liberalized. Four years later the protectionists get elected and the rules change.

Do not be surprised if a developing country still has a macho culture. Females may not be in positions of authority outside the government. If you or one of your imported American engineers is a female, expect difficulties in dealing with contractors and suppliers on an equal footing. Expect to be hit upon or even touched in social situations. Making it obvious that you are deeply offended will make your job even harder. Watch how the local women handle it. Ignore it, dress very conservatively and remain professional or go home. You can't change their culture yourself.

Other developing countries, especially where Islam is the predominate religion, will have other restrictions on women. Head scarves, conservative tops, and long pants or skirts are the norm even at the beach. Women may not be able to drive cars or even go about without a male escort.

I was in South America during the President Clinton impeachment and Monica Lewinski scandal. The locals thought it was funny. To them it was an indication of how uptight Americans were. They said in their countries their politicians all did the same thing and no one paid it much attention. Some politicians even bragged about their sexual exploits to garner the macho vote. They said Clinton's mistake was getting caught and then lying about it.

Many Islamic countries severely restrict the sale and consumption of alcoholic beverages. They may be available in hotels frequented by foreigners but seldom elsewhere.

If you have an allergy to cigarette smoke you better bring some pills. Smoking is widespread. There may be, in hotels catering to Americans, smoke free areas of restaurants. Your hotel rooms probably were smoked in. Expect the residue to be very apparent.

With all these restrictions why would a company elect to build a facility in a developing country? Access to raw materials and very low relative labor rates are two good reasons. Expanding their brands world wide is another. So is cheap energy. Even with the restrictions, expansion into the third world can be very profitable. And any rule or regulation is always negotiable.

CONSTRUCTION IN THE DEVELOPING WORLD

If you believe that you are going to obtain the same quality and type of construction you can obtain in the developed world you are in for a surprise. Here we have to separate third world from second. In the second there may well be precast concrete wall and roof panels available but you have to watch the quality. There may be good metal buildings available, but again, be quality conscientious.

A lot of the world builds by constructing a reinforced cast in place concrete building frame, usually with cast in place concrete floors and roofs. Reinforcing is adequate but if you think you will see neat carefully tied reinforcing steel cages and mats, think again. It is not haphazard but it looks that way. Once the forms are off some steel may be visible on the surface. They cover it with stucco to hide it.

They then fill in the space between the frame members with a form of structural clay tile blocks which are honeycombed to save weight. These structural tiles are placed in mortar and the interior and exterior are covered with stucco – at least they are stucco in nicer buildings. They don't bother in cheap buildings. It makes a reasonable wall and is easy to modify.

Since this tile fill has no real structural value besides holding itself up, knocking a few tiles out to install a window or door or run a pipe is easy. All but the largest window and door frames are their own lintels. I have seen them place structural tile without out regard to any penetrations and then come back a week later to knock out portions for windows and doors. It creates some insulation value with all the air trapped in the honeycombs and it is very resistant to wood destroying organisms like mildew or insects. But it creates a harborage for insects and vermin, also.

Unless you carefully specify a gravel substrate and poured and reinforced concrete for floors on grade, you can get a poor mixture of dirt and concrete for a base layer, and then an inch or two of concrete on top. This does not hold up to much more than foot traffic. In other areas you may get a hand laid rock mosaic like substrate with concrete on top. This is even better than our system and holds up very well if the topping is a couple of inches thick.

In some areas the contractors are paid weekly based on the volume of materials excavated or placed. They then pay their workers the same way. It requires a large group of people involved in taking measurements. You may see an area being excavated that has one or two unexcavated pillars in the center. These pillars establish the top of the existing grade prior to the commencement of excavation. It is a base line to calculate the cubic meters of material moved. The last parts excavated are the pillars.

Expect contractors to want paid upfront for materials. Few have the internal financing to purchase high volumes of materials and then wait for payment. Insisting that developers and foreign firms make a down payment when a contract is issued is a normal business practice.

As said previously, almost everywhere outside the USA uses the metric system. Yet so much material is imported from the United States, especially in Central and South America that our standard pipe sizes, wire gauges, and bolts, nuts, and fasteners are there. Material becomes a mix of English and metric. Buildings will be constructed to metric measurements. Pipe will be US standard pipe unless sourced from locals who may or may not provide metric. Be prepared for a load of three inch pipe to be a half load of US standard three inch and a half load of 90mm. Although nominally the equivalent, US 3" pipe is 88.9mm OD. The closest metric equivalent is 90mm OD. They won't fit together and when welded will leave a bump.

Don't expect the quality of welding that you get in the USA. If you don't insist on quality you will get skips and voids. Inert gas welding for stainless, aluminum and other exotic metals is not universally available and they will try to stick weld these materials. It isn't pretty. If you need automotive or truck mechanics you will be pleasantly surprised. They have been keeping vehicles on the road that we long ago would have regulated to the scrap pile. And they can even make their own parts and often have to.

Electrically, most developing countries sure don't follow the US National Electrical Code and don't use much conduit. Wire is rubber insulated cable and is placed in cable trays and then dropped to equipment locations. A lot is cord and plug connected using the plug as the disconnect. Communication cable is tie wrapped or taped to whatever and draped from point to point, unless you get real specific. Ground faults are normal, grounding is less than adequate, and many over current safety devices are either not there, are of insufficient interruption capacity, or don't function. Power isn't on more than 80% of the time. In many areas expect to install your own backup electrical power generation capability.

Broadband electronic communications can be quite limited. Connections are slow and intermittent. Land line telephones are noisy. This isn't a surprise. If you look up in any major city or town at the aerial spaghetti tangled from telephone pole to telephone pole to buildings without any apparent rhyme or reason you will see why. Cross talk on land lines is typical and you may get to listen to two or three conversations all going on in the background during your calls.

There are some areas with excellent cell phone communications but your USA cell phone will usually not work. The frequencies are different and the protocols don't match. Get a local phone and SIM card. You may even be able to get some 3G or even 4G coverage. Cell phones are cheap, local calls are reasonable and even overseas calls aren't bad if you go through ATT's USA direct operator. But 3-4G direct internet may be expensive and a Blackberry or Iphone may not work for internet even if a cell laptop modem does. Every country and area in the country is different. If you are really in the boonies, Iridium satellite phones work well, but air time is expensive. Use the portable dish for data transfer. It is far more reliable than the phone whip.

You may find that you need to set up your own Wi-Fi hot site (secure) and even lease a phone line to get your email. In the interim, you can use internet cafes that are in most major cities. Major hotel chains may have Wi-Fi and most have Ethernet connections in their rooms, but check price first. Some are very expensive per unit of time, as are overseas telephone calls from hotels. If they use a satellite link they charge by the data packet. A twenty minute phone call back to the USA may cost you $150. This is robbery, so use a cell phone for voice and internet if you can.

Developing world contractors do not provide the level of supervision we are used to for their workers. Workers will build a wall or run a pipe the way they always have. If it isn't being done the way you want it, you have to communicate. If you require a contractor to tear out something that was installed the way they have always done it and do it your way it is an extra and you will be expected to pay for the extra. The key is detailed drawings and specifications in their language followed up with meetings to bring the difference to their way of doing things to their attention. Don't skip the meetings. They usually don't read specifications and they see what they expect to see on drawings.

Normal contractor work days are six days a week, ten hour days. The days may vary from Sunday through Friday or Monday through Saturday. There are normally a lot more holidays at different times of the year. Contractors do not work on holidays unless paid very well for it. There is no work on religious holidays even if you wave lots of money. In countries like Brazil, Carnival (just prior to Lent) is almost a religion in itself and all factories shutdown for the week. A national soccer championship game will stop everything for at least a day.

Years ago I landed at Maiquetia airport serving Caracas just as a local "football" championship game was starting. We stood in an unmoving line at immigration for three hours until it was over. All the agents were watching the match.

Strikes or slow downs can be frustrating. Usually it is government worker unions. Railroads or buses just stop or go into turtle mode. Once I spent ten hours in Guarulhos Airport in Sao Paul, Brazil, in an enormous line trying to get through security for a flight to the USA. The security guards were opening and slowly inspecting every carryon bag and patting down every passenger. They managed to mess up the schedules of airlines and passengers all over the world as flights just sat waiting for passengers. I think the slowdown was about a less than desired salary increase.

Weather conditions drive work and weather extremes can be significant. In the tropics, some contractors do not work between 1PM and 3PM. To get in the ten hour days, work starts at 6AM, goes to 1PM, stops for lunch and siesta, starts again at 3PM and goes to 6PM. Each country, and even area in a country is different. It is usually driven by the hours of daylight.

Torrential tropical rains can pop up and then stop, leaving puddles and the ground steaming. An hour later the puddles are gone. Humidity is significant. In the tropics the day time temperature can easily top 100 degrees F, in the summer, and 85 degrees F in the winter. But usually there is no season defined as summer or winter – just dry season and rainy season.

SAFETY

It is safer then you think but not any safer than it would be in a major US city, especially if you wander around in poor areas. Most developing countries have police that play hardball with petty criminals. Tourists or foreign business people that get robbed or worse, generate bad press for the government. In rural areas you will find that doors, if they even have doors or windows, have no locks. The residents basically trust each other.

Police can be federal, regional, municipal, private or all four. The federal police are usually part of the military and are well trained and ruthless in dealing with criminals. Regional and especially municipal police can be interested in a little side money and may well pull over driving foreigners looking for a little tip prior to letting them go on their way. They will want to see your papers and the car registration card. A twenty dollar bill in your passport when you hand it over does wonders to insure that you are quickly allowed to proceed.

If they take offence at a potential bribe just play dumb, apologize and take the $20 back; playing dumb in these types of situations works well. Even if you speak the language, pretend not to totally understand. They think foreigners are stupid anyway, so you will fit right in. Be 100% cooperative with police. They take a very dim view of anyone who even remotely challenges their authority.

A note on passports: most countries outside the USA require their citizens to have and carry a National Identity Card. Without it a person can't cash a check, use a credit card, register at a hotel, buy a train or airplane ticket, get a job, or even drive a car with a license. It has their photo, finger prints, name, address, and citizen status on it. The locals guard their identity card like it was a very valuable possession – which it is. Without it they are non persons. There are few illegal aliens. They can't do anything without a National Identity Card and if caught, they are incarcerated or deported immediately. That's the good news. In some countries, the good news happens after the bad – being beaten senseless by the local police.

Your identity card is your passport. You will need it to do any of the things that the locals need their National Identity Card for. Keep it, or better yet a color Xerox copy of it including a copy of your entrance visa and stamp with you at all times. Also, keep a copy in your suitcase or room safe at your hotel. If you lose your passport you are in trouble. But with the copy you can quickly get a replacement at the US Embassy or Consulate. Many hotels will require that you give your passport to them for the duration of your stay. Then you must have a copy to keep on your person.

Most developing countries will not allow foreigners to work inside their country without a fancy work visa. However most interpret this to mean foreigners that are there for an extended period and who do not periodically go home. So, if you go there for three weeks every six weeks or so, they will consider you to be a tourist and leave you alone. You may need a special multi-trip tourist visa.

In many countries, if you must get a work visa and you receive your pay locally, they will expect you to pay income tax on the money you made while working there and may insist that it be paid prior to allowing you to get on an airplane home for a visit. Keep records. Tax paid to a foreign country is usually creditable against any US income tax owed. Most ex-pats get paid both in their home country and in the work country and the amount in the work country is selected to just come in under the local tax rate step. Of course if you can choose a home country with zero income tax like Singapore, it is a good deal.

Many countries will not allow a woman to take a child out of the country alone regardless of her marital status. This creates a surprise for Americans who have a family and who relocate to the country on a temporary basis bringing their family with them and the wife wants to take the kids home for a visit. Check with the US Consulate. You may need a notarized letter from the husband authorizing such travel.

As a matter of course, the local rich live behind guarded walls, use bodyguards, and their vehicles are bullet proof. This is primarily because kidnapping the rich is a thriving and lucrative business.

Every major city has pickpockets. Some are very good. They work in teams and do the sandwich trick where one stops abruptly in front of you while another bumps into the back of you. Each apologizes profusely. Later you notice that your wallet, watch, etc. are gone. One group was composed of three young women in tiny sexy outfits that preyed on foreign men. The third girl was the lookout. They carried box cutters in case their marks reacted prior to their getaway.

Keep your wallet in your front pocket preferably with your hand on it. I can remember once during a festival in a crowd, one of my engineers said he was walking down the street with hands in both his pockets – and neither of them were his.

Many countries, especially in Central America have drug entrepreneurs and various groups compete for the lucrative business. This competition is war and innocents do get caught in the crossfire. Stay clear.

Common sense should prevail. If you wouldn't go wandering around alone in the USA in an equivalent area, don't do it in a developing country. Take the minimum funds you need, dress simply, leave your jewelry at home, and be prepared to gladly give up your pocket cash if accosted. Let the police be the heroes.

ILLNESSES

Do not be surprised if any people you import from North America or Europe get deathly sick the first week or two. Travelers to Mexico will know this as Montezuma's Revenge. It is a gastro intestinal malady roughly the equivalent of radiation poisoning. Those afflicted will try to turn themselves inside out to expel the offending invader. The food contains different bacteria than we are used to, even salmonella, and the locals are used to it and it doesn't bother them. In most cases this illness will go away by itself in three or four days. Bed rest and lots of fluids to maintain hydration are required. Coconut water is the home remedy recommended by the locals. If it lasts longer than five days, you have something else more serious. Go home and see a doctor.

Once your system has had the opportunity to adapt to the local bacteriological environment, usually two to three weeks, you are fairly immune to most forms of the "revenge". But if you go home and don't come back for six months, expect to get to adapt again.

Initially, I would recommend staying away from salads and don't drink the tap water. Even if a salad looks clean and fresh it was undoubtedly washed in tap water. Fruit is generally safe if you can peel it yourself. Raw shell fish are very suspect and should never be ingested. Any meat you eat needs to be well cooked – especially pork, chicken, goat and rabbit. If you cut into a piece of chicken and see pink don't eat it like you might at home. Fish, especially salt water fish, are great if well cooked and flakey. After a few weeks you can get a bit adventuresome and try some of the local salads and other dishes. Most are wonderful.

I love sushi having lived in Japan for three years a long time ago. But I don't eat it in developing countries. It is like playing Russian roulette only with five bullets and one empty chamber in the six gun.

Watch for the small bowls of sauce on the tables. In most countries that serve spicy food it is these small bowls that contain the real spice. And some would power a moon rocket. Gingerly taste a sauce prior to putting gobs on your food. In South America, especially Brazil and northern Argentina, you will find a bowl of what looks like uncooked cream of wheat on the table. It is manioc root and has no discernible taste to me. The locals use it a lot on almost everything – especially the staple side dish of beans and rice. You will also find that there are hundreds of varieties of plantains and potatoes. If you ask for potatoes you almost always get French fries.

Eggs will not taste the same. Their chickens don't eat what our chickens eat. To me their eggs are gamey. You will also find that they eat bacon but do not cook it anywhere near as much as we do. In fact, most developing world countries do not have our obsession towards limiting the fat in their diets. Yet few are overweight. I believe it is because most get a lot more exercise than we do, three square meals a day are only for the rich, and portions are a bit smaller.

Desserts are usually super. And super charged sources of calories. Sugar is world price sugar, very cheap by US standards and used accordingly.

You will be surprised. There are some very nice restaurants, some that would easily be considered worthy of five stars in the USA. They will be only in the more developed areas. They will be crowded by 10:30 PM, full of the rich members of the population and will not be cheap. In Central and South America don't expect dinner at 6:00 PM. Most restaurants don't even open until 8:00 or 8:30 PM. At 9:00 PM you may be the only patron.

If you only stay in nice hotels, the water is filtered and safe there. This is important since one of the problems with water is a tiny parasite called Guardia. It causes lower tract distress. It doesn't seem to bother the locals as much, but will make outsiders very sick and is fairly difficult to get rid of. And it can take a week for symptoms to show up. You can also contract this parasite if you swim in fresh water lakes or river pools. You can contract a lot of other parasites doing this and it is not recommended.

Watch out for bottled water. If it is not sealed (the cap should be attached to a breakaway ring) and sold in a reputable supermarket or restaurant it is suspect. I have seen a ramshackle roadside facility refilling empty plastic water bottles from a public spigot and sealing the bottles with a used cap and a strip of clear plastic tape. And ice is no safer than the water it is made from. Bottled beer is pasteurized during the bottling process and most is excellent. Milk can be suspect if not pasteurized. The same is true of cheese, especially the softer cheeses.

There are documented cases of foreigners getting very sick. I knew of a man from Australia who picked up an amoeba in Brazil. He guessed that it was from ice from a beachside tiki bar. And a shot of rum in a glass of juice and ice is insufficient for an amoeba kill. He was hospitalized for months and lost 30% of his body weight. He was on restricted duty for six months after he got out of the hospital. The best answer for anyone who gets very sick on assignment to a developing world country is a visit to a local doctor for stabilization followed by a one way ticket home. The level and sophistication of medical care in the developing world doesn't match ours. One company I worked for had an American tropical disease specialist on retainer. His office was in Maryland. We didn't use him often but he was there when essential.

The developing world has a lot of older diseases we have eradicated. Pre trip inoculations are essential. Tuberculosis is there. Mosquito born illnesses are there. Dengue fever kills thousands in tropical Central and South America each year. Yellow Fever, Malaria, Encephalitis, West Nile Virus are there. Bring an insect repellant and use it. Coastal areas, especially where there is always a trade wind don't have flying insects. They get blown inland. But if you go inland or into areas protected from the trade winds, expect to find them. They range from tiny bugs (called "no-see-ums" in the southern USA) to huge mosquitoes that could mate with a 747.

Hepatitis is there in all its configurations. A is the widest spread. There are a lot of unsavory maladies in the developing world, and you should insure that you and your imported troops all get whatever the Center for Disease Control recommends for the countries you will be visiting and keep these efforts up to date with booster shots and pills.

Sanitation is hit or miss in some areas. Major metro areas are usually well served, but the moment you get into anything approaching a rural or slum area, conditions deteriorate rapidly. Roadside ditches are also sewers. It can get a little ripe on a hot day. Usually only newer developed areas have anything that even approaches waste water collection and treatment.

Beaches may look inviting but the fecal chloroform content of the water on some of them would cause an immediate beach closure in the developed world. Only swim on beaches that are monitored and supervised. Watch for shark warnings. Most good beaches have shark nets where sharks have historically been active.

Watch for the effects of the tropical sun. Light complexion people can get burned in 15 minutes. In tropical areas the sun is directly overhead most of the day and strong. Use sun block! Most of the locals have darker complexions and can take it. Women's bathing suits are no larger than postage stamps in Brazil and men's are not much larger. They expose far more skin that wouldn't be exposed in the USA. If you dare to wear one, this fresh meat will fry quickly.

Trash pickup can be hit or miss and piles of garbage covered with vermin pop up regularly. Cockroaches are prevalent and big enough to throw a saddle on. Then there are the multiple varieties of snakes – many venomous. There are also spiders and in some areas scorpions. If you are going to go tromping through fields or underbrush do it with heavy pants tucked into high top boots.

Many areas have lizards. Don't hurt the small ones. They eat their weight in insects each day. If you have one in your hotel room it is considered to be a sign of good luck. Leave him a saucer of water before you go off to work. If you see big ones over a foot long in the bush leave them alone. They are aggressive and their bite is toxic.

Watch out for your people self medicating. Most developing countries do not have anywhere near the restrictions we are used to as to what prescription drugs can be purchased over the counter in local pharmacies. Strong antibiotics, stimulants, depressants and narcotic drugs can be had far below the US price and without any warnings. Mixing some of these with alcohol can have disastrous effects.

There is also opium and its derivatives and cannabis. Getting caught with these, or worst case getting caught trying to smuggle some back to the USA, gives the perpetrator an opportunity to experience the local justice and incarceration system. It is something out of Dante's Inferno. If you have ever watched an episode of "Locked up Abroad" on the National Geographic Channel you know that this is very bad news.

Executing a project in the developing world can be a challenge. Never assume that your expectations for the quality or materials, labor, supervision, and timely responses that you receive in the developed world apply. Be health and security conscious. Learn their culture and accept their norms. Enjoy the differences and don't judge or try to Americanize the people.

You will learn a lot, come to appreciate another culture, and make a difference in the lives of many. I made a lot of friends in far distant places. You will also.

Section IV, Chapter II

TECHNICAL DEVELOPMENT

Many industries get by with purchasing off the shelf equipment to execute the production of their product. Go to a trade show and look, contract for a turn key line from a reputable manufacturer or contract with a system integrator, and get your new production system.

The next week your competitor will do the same and guess what? His or her new line will look just like yours, use the same number of operators, produce the same levels of scrap, and even make a similar appearing product at the same speed and quality. There is very little manufacturing competitive advantage in that.

If you are a small business with limited resources at your disposal you may just have to do it that way. If you have the advantage of more resources you may want to think of developing some of your own specialized equipment to achieve a competitive advantage.

There is another force that may drive you to technical development. A new product that is so unique that it just doesn't fit the commercially available equipment or production lines. Here you are forced to develop your own equipment.

Technical development is expensive – very expensive. One just can't whip out a few drawings, fabricate and assemble some parts, hook up some wires and have a new machine ready to run reliably. I wish it were that simple.

The first overriding question you must ask is technical development necessary? Carefully evaluate what is available commercially. Can you modify a piece of off-the-shelf equipment to do what you need without starting from a clean sheet of paper? Check out equipment used by industries other than your own. Could some of their stuff be modified to handle your product?

Trade shows are great for this. I would send a small team of engineers to some of the larger trade shows, each with a specific objective to research. Who has some new equipment offerings? What are they claiming in throughput? What new technology are they touting? How does their new equipment compare to ours, to their competitors? Consider sending a team to a trade show for a totally different industry to see if there is something you can leverage.

Many in upper management consider engineers attendance at trade shows as a boondoggle: a chance for engineers to wine and dine, stay out late and sleep late, and generally goof off at the company's expense. There could be some of that, but a trick is to show up yourself the last day of the show and have each engineer you sent take you to booths where there was something significant to see. This technique focuses their minds and will allow you to report to upper management on the significance of what was seen and the importance of attending trade shows.

If you have determined that you can significantly modify commercially available equipment to meet your needs, you have to acquire it and commence modification. You need to be very careful. Equipment manufacturers are not stupid and can sense that you may be just looking for something to buy and modify. In some industries there are companies that do this repeatedly and equipment manufacturers know and may be reluctant to sell you one of their new machines.

If you tell them you want to buy one just to test it out on your line, they will want to participate in the test. Best case is to buy one and say you will be testing it on a new product and that no, you don't need their assistance. Later when they call, and they will call, say the new product was dropped from consideration.

Never buy one machine and then expect to duplicate it multiple times in-house. It is a patent violation and you will get hit for it the moment the manufacturer finds out.

Remember, the moment you make the first modification without involving the equipment manufacturer you must terminate your involvement with the equipment manufacturer and accept that you have voided any warrantee or performance guarantee.

You may still purchase spare parts for the original equipment but it is no longer their equipment and under no circumstance should anyone, salesman, technician, anyone from the manufacturer or any systems integrator working with the manufacturer ever be permitted to see the equipment again. You paid for it and it is yours.

In extreme cases, you could be charged with violating their patent if you use some patented part of their machine without involving the manufacturer. This is a fine line and most equipment manufacturers will not get uptight if you refine their design a little to make it a bit faster or more reliable, especially if you share your changes with them. But if you utilize the technology they developed and patented for other things without telling them you have a potential for a law suite.

Most companies involved with the conversion of existing equipment to do other things do it in total secrecy. They have to change the look and feel of another manufacturer's equipment sufficient to insure that it no longer looks like or infringes upon the patent of another manufacturer's equipment. Never even consider taking a piece of equipment you have modified, duplicating it and selling it on the open market. Internal use only has to be the rule.

Patent law is currently in flux. In the past a technology was not patentable but a piece of equipment using that technology for a specific task was. This distinction is getting fuzzy. Some technology itself is getting patent protection. Be careful.

Should you patent equipment you develop for your internal use? Maybe. The patent process is long and cumbersome, the patent office leaks like a sieve, and your ideas will be in public view just when you want them to remain a secret from competitors. Other patent holders may challenge your application to either obtain royalties or just to get a better look at your equipment.

Your business has its own guidelines for intellectual property. Find out what they are.

If you can't find anything you can modify in the realm of commercially available equipment you are faced with starting from scratch. This is the most expensive form of technical development, takes the longest but potentially has the highest rewards. Try to fairly assess the cost benefit ratio. If successful, just how much of a manufacturing competitive advantage will the new development deliver? If the idea isn't going to make a significant change in how you manufacture your product, is it really worth all the cost and time it will take?

Assess the underling idea carefully. Converting a process that currently uses intermittent motion to a smooth rotary motion usually is a good move. Going from a batch process to a continuous one reduces variables and improves quality. Employing high speed servos to precisely control machine motion almost always allows for higher speeds and more flexibility.

Then staff your technical development skunk works with a few very creative people who share the vision of success. Insure they have the resources they need including not only thinkers but doers – technicians, mechanics, machinists, etc. They should be able to make prototypes themselves to test concepts.

Do not expect them to report often. Remember that the more you have to report on what you are doing the less time you have to do anything. Stability is achieved when you spend all your time reporting on the nothing you are doing.

Expect the development to be an iterative process: idea, prototype, test, repeat. Provide guidance and just stop in to see what is going on now and then. Sometimes just slip resources under the door and let the group function without you breathing down their neck all the time.

I fully understand that this is difficult to do since for any significant technical development you had to get the money and resources approved from upper management, and they will always be on your case about delivery. When do we get it? Why is it taking so long? Why does it cost so much?

Most significant technical developments I have been involved with take years to perfect. And do not be afraid to shut a development down if it appears that there is little progress over the long term. You may have overreached the capabilities of technology, have the wrong people assigned to the effort, or a commercial equipment manufacturer may have just announced a new machine that will do 95% of what you were trying to achieve. Commercial manufacturers have their own skunk works and sometimes have more expertise and resources than you do.

There are outside engineering firms that specialize in technical development for corporations. There aren't many and they are quite secretive. If you visit their facilities you get to walk down lots of halls with curtains on each side and "Restricted Area" signs.

These firms perform a valuable function. Usually they are staffed with some extremely creative and technically savvy people and wouldn't still be in business if they had not delivered the technical developments that their customers wanted. They are very expensive. If you use one or more of these firms make absolutely certain that your contract with them delineates who owns the designs and prototypes along with the final machine. You want to be the sole owner.

Too often I have seen companies hire one or more small engineering firms to develop and build a special machine for them only to find out that the small engineering firm has sold the machine or one very similar to a competitor. If your development becomes more than 20% of the small firm's business, once they are done with your work do not be surprised if they try to leverage the knowledge they developed or obtained from you to replace the income. In fact be very careful that you do not become a critical portion of any supplier's total income. They become dependant and exceedingly desperate if you stop feeding them, and they will do anything to survive.

Many companies contract with universities to do some technical development for them. Universities are cheaper since all the grunt work is usually performed by graduate students. Don't expect total secrecy. Professors publish or perish. Graduate students write theses. Universities are a lot better at developing technical concepts than they are at actual functioning machines. If you need a concept fleshed out and tested, they are great at that.

Many companies establish an ongoing relationship with two or three universities. Then either the company can propose work, or the university can propose chasing an idea that may be applicable to the company. These ongoing relationships can, and usually do, require a sustainable commitment of cash flow to one or more engineering programs at the university. Hint: each university should have one manager at your company that is their primary contact and who visits and meets with the professor in charge of the university's program on a periodic basis.

If you need some pure research into a technology, recognized development laboratories both commercial and government can help. Yes, government research laboratories take on development work for private industry. It keeps the cash flowing. Your development work will not have the priority if a new government assignment comes in and there is little or no secrecy here, but government development labs do have an enormous talent pool and can be of value. Don't expect a lot of urgency to get your work done.

Of the technical development work I have been lucky, or unlucky enough to be involved in over the years, a few were enormous successes which drastically reduced the company's manufacturing cost and gave us a significant competitive advantage. Major feather in cap. A few were break even – what we spent on the development and what we saved were a wash. And a few were expensive utter disasters. It is just the nature of technical development.

-The End.-

Jack L. Wells

Engineering Management, an Irreverent Primer

Jack L. Wells

Other Books by Jack Wells
All novels

Jack Wells' first book
Published December 2007

QUICKSILVER: a greyhound at sea

CDR Jack L. Wells, USN Ret.

A HISTORICAL NOVEL OF THE
MEN WHO FOUGHT THE VIETNAM WAR AT SEA

ISBN 0-7414-4059-8

This book is the story of the officers and men serving aboard a fictional destroyer in combat during the Vietnam War. It follows the experiences of one young naval officer ENS, later LTJG Patrick Dillan, USN.

QUICKSILVER: a greyhound at sea.

Overall Rating: ★★★ **Three Stars: Recommended.**
A solid effort. Commander Wells writes with authority which makes this fictional account of a destroyer's action during the Viet Nam war very readable... The characters come alive and most readers will find themselves putting names and faces to each from their own experiences... The day to day life at sea is familiar to any who has been there...and one can almost feel the sea beneath them as they read. The detail on every page will keep the reader attached. The author's credibility is established early on and continues to the end... Two or three story lines are too important to reveal in a review and the readers will have to find them for themselves. They add personal and tragic dimensions to this well written novel about a time that many might like to forget.

Bernie Ditter
Reviewer,
The National Association of Destroyer Veterans

Thoroughly enjoyed **Quicksilver: A greyhound at sea**, and I've asked our manager to order copies to sell in our gift shops. It seems only natural that this book should be available in our book selection. We were young then, and I'd forgotten how exhausted we were most of the time...

William Metcalf
Reviewer, Executive Director
Historic Naval Ships Association

Few readers today know what the Navy did off the Vietnamese coast in the 1960's, but *"Quicksilver: a greyhound at sea"* will quickly educate them…Author Jack Wells, a retired Navy Commander, has written a historical fiction novel based on his time at sea during the Vietnam War. He's placed a crew of officers and enlisted men on a Gearing Class Destroyer and given the reader a front-row seat of their year away from home. Although Vietnam was primarily a land war, the Navy conducted multiple missions, and Wells brings both the tedium of sea duty interspersed with the exhilaration of combat…This is a well-written book. A former mustang officer, Wells has an eye for detail and writes from the viewpoint of one who has "been there." His characters come alive and are believable; no small feat for a first-time author. With so few books on the market about Navy actions in Vietnam, *"Quicksilver: a greyhound at sea"* would have made an excellent memoir of a time, place, and actions that few people know occurred…**Recommended !**

Prof. Andrew Lubin
Reviewer,
**Military Writer's
Society of America.**

Quicksilver: a greyhound at sea: You have captured incredible detail of real life on a greyhound of the sea. Your story of young men driven to demonstrate flawless decision making while stressed by fatigue and tension could not be more accurate. I thoroughly enjoyed how you developed characters to convey life at sea and how you showed the importance of supporting your men… It ought to be required reading for all naval officer candidates.

Virg Erwin,
Author, *Cat Lo, a Memoir of Invincible Youth.*
MWSA Gold Medal winner 2009

Jack Wells' second book
Published December, 2009

PAPER DRAGON, WOODEN SHIP

CDR Jack L. Wells, USN (Ret)

BRONZE MEDAL AWARD

PAPER DRAGON, WOODEN SHIP

210

A HISTORICAL NOVEL OF THE
NAVY SET IN THE FAR EAST -
A STORY OF THE THE COLD WAR
AND THE VIETNAM WAR AT SEA

ISBN 0-7414-5240-5

Paper Dragon, Wooden ship continues the story of Patrick Dillan introduced in the first book: *Quicksilver: a Greyhound at Sea.*

In the second book Patrick is transferred to the USS BANNER (AGER-1), the sister ship of the ill fated USS PUEBLO (AGER-2), and is home ported in Yokosuka, Japan. He gets sent off to Survival Evasion Resistance and Escape

(SERE) training and conducts intelligence operations in the Sea of Japan.

When BANNER is decommissioned, he is transferred to the staff of Commander, Mine Flotilla ONE in Sasebo, Japan. The Mineflot period involves Market Time combat operations off Vietnam and later riverine warfare in Vietnam.

He also goes through a relationship failure, not untypical for marriages in the military, and the establishment of a new relationship.

Reviewed by
Military Writers Society of America

Title: **Paper Dragon, Wooden Ship**
Author: **CDR Jack L. Wells, USN (Ret)**
ISBN: **0-7414-5240-5**
Publisher: **Infinity**
Genre: **Historical Fiction, Navy**
Reviewer: **Joyce Faulkner**

Like many military memoirs, *Paper Dragon, Wooden Ship* follows a year in the life of a Naval Officer. Pat Dillan arrives in Sasebo, Japan in 1969, shortly after the North Korean's captured an intelligence ship, *The Pueblo*. He is assigned to her sister ship, *The Banner*. His life is complicated by the decay of his marriage and the changing political perspectives of the times. When his wife leaves him during a trip back home, he returns to Sasebo to a new rank, new assignment and a new love interest. It's a familiar story written with a twist – it's written in Navy, not English – and it's a novel, not a memoir.

This story is unusual in that both men and women will find it romantic and intense. There's action, political intrigue, nostalgia for another time and place, the relationship a man, a woman, the Navy, and war – both hot and cold. The reader can almost hear Mick Jagger singing, "You Can't Always Get What You Want" in the background as Pat struggles to balance his personal and professional lives. The author allows his characters to define themselves through their words and deeds. Their conversations are real, amusing, and convoluted – just like everyday folks. CDR Wells peoples his novel with likeable Americans and allows the villains of the time to create conflict for the good guys. Most interestingly, the Navy seems more like a character than an institution – warm mother, strict father, petulant lover, demanding professor, intrusive in-law all wrapped up into one. It's a daring ploy by the author, but in the end, it creates an intriguing and unusual story.

The cover supports the author's intent with the look and feel of a non-fiction publication – with a faded background photograph of a Japanese pagoda and silhouettes of Pat's two ships in the foreground. This matter-of-fact approach makes this historical novel seem more real than most – like the personal story of your next door neighbor -- the super-intelligent one that speaks in a tangled dialect of alphabet soup and hides his heart behind a short haircut and shiny black shoes.

While the plot is compelling and the characters intriguing, this book is not an easy read. … To be fair, CDR Wells provides lots of footnotes and goes the extra mile with an Appendix to help decipher Naval ranks. However, most of us landlubbers may spend as much time looking up terminology like EOD, Mike Boat, and LCM-6 than actually enjoying Pat's adventures in war and love. However, this book will enchant those who live and love the Navy – and after about fifty pages, a newbie will grasp the lingo enough to get a kick out of it too.

Joyce Faulkner, President, MWSA

Paper Dragon, Wooden Ship was awarded a **Bronze Medal** in the Military/Navy category by the Military Writers Society of America

Jack Wells third book
Published February 2010

Jack L. Wells' novel the *Breath of the Choson Dragon* was awarded the **Gold Medal** in the Mystery/Thriller category for the year 2010 by the prestigious Military Writers Society of America. It was also awarded a **Silver Medal** in the 2010 Branson Veteran's Week Stars and Flags Book Awards in the Fiction category.

Military Writers Society of America REVIEW: *Breath of the Choson Dragon* is a high energy, well researched, completely plausible scenario of patient, but fanatic war planning by North Korea. A gripping tale of their desire to reunite the Korean peninsula, the north is shown, accurately, to be an enemy that is not to be underestimated, or taken lightly. Their quest, against staggering odds, to deal a severe surprise nuclear blow to the United States, and Japan via a plot that spans decades of tedious preparation, at levels so secret as to be almost impossible, is told with meticulous detail by the author, an ex naval officer, who has been there, and done that. So many military books are written without the benefit of firsthand experience that they bog down in the minutia that is second nature to veterans. This one feeds voraciously on those details that authenticate the book as the real deal. This author writes, and thinks with the no time for nonsense frame of mind that is necessary in the military when it is faced with clear danger.

As fascinating as the realities of submarine warfare, and nuclear weapons systems are, the story line that weaves in and out of those technical revelations are just as compelling. The tech, and human interest aspects dance in tune, step for step. Neither Clancy, nor Ludlum have written anything better, and the author, Jack Wells, is their equal, at least. A fast paced story (a movie producer's dream) is fed by the intriguing reality of our nation's military, and intelligence agencies' daily challenges, told with the voices of real human beings that the reader can identify with. The political dilemmas of the main characters, which come hand in hand with their highly secret jobs, give this yarn dimensions that are consternating, frustrating, and agitating, just as they would be in the real world. The ending is clean, and satisfying. I wanted to cheer. I felt good about it. I choose to believe that if this, or a like scenario ever came to pass in reality, our soldiers, sailors, and airmen, would receive the support they needed from their political masters to affect the same conclusion. Nowhere in the human experience is teamwork, trust, diligence to detail, and precise behavior as required as it is in the armed forces when lives are at stake. Thanks to this author, readers who did not serve in such capacities are offered the opportunity to appreciate these men, and women.

Review by Bob Flournoy, MWSA Reviewer (May 2010)

www.jackwellsauthor.com

Jack's first and second books were published by Infinity Publishing.
You can purchase Jack Wells first and second book through www.buybooksontheweb.com
or by telephone 1-(877) 289-2665 for $19.95
They are also available
through the Military Writers Society of America
www.militarywriters.com

Jack's third book was published in February 2010 and is available through the Military Writers Society of America
for $21.95
www.militarywriters.com

These three books are also available through the largest online bookstores. Search by title, author or ISBN

You may also obtain a signed copy direct from the author and publisher for the cover price plus $5.00 shipping:
Contact the author through the author's web site
www.jackwellsauthor.com

Sextant Publishing
a division of Wells Inspections, LLC

www.ingramcontent.com/pod-product-compliance
Lightning Source LLC
Chambersburg PA
CBHW071359170526
45165CB00001B/111